FROM MALIN HEAD TO MIZEN HEAD

FROM MALIN HEAD TO MIZEN HEAD

A JOURNEY AROUND THE SEA AREA FORECAST

JOANNA DONNELLY

Gill Books

Gill Books
Hume Avenue
Park West
Dublin 12

www.gillbooks.ie

Gill Books is an imprint of M.H. Gill and Co.

9780717197361

Designed by Bartek Janczak
Print origination by Studio 10 Design
Edited by Gráinne Treanor
Proofread by Jane Rogers
Printed by Printed and bound in Great Britain by CPI
Group (UK) Ltd, Croydon, CR0 4YY

This book is typeset in Adobe Garamond Pro.

*The paper used in this book comes from the wood pulp of
sustainably managed forests.*

A CIP catalogue record for this book is available from the
British Library.

5 4 3 2 1

CONTENTS

This book is dedicated to the memory
of my mother, Marie Kelly.
Her sudden departure from this world
set me adrift. This journey around the
headlands has helped me to start to
right the course.

INTRODUCTION

I'm often asked – in fact, it's a guaranteed question – what, for me, is the favourite part of my job.

People grow up with the weather forecast. It's on in their houses because grown-ups want to watch the news. The weather forecast comes on after the news and it's something that kids are okay with. It's the same every day – basically. Yes, the weather is different, but the presence of a presenter standing in front of the blue screen is the same. No deaths, births or marriages, a fixed length and format. It's nice and reassuring, and kids like that.

So, kids grow up and they see these guys presenting the weather and they think, that's a weather forecaster. Some think they'd like to *be* a weather forecaster, some even think they *could* be a weather forecaster.

I was not one of those people. I was only interested in maths or science and absolutely not in weather. Don't get me wrong, I've always loved a good thunderstorm, but I live in Ireland, and we don't get good thunderstorms here. I've seen more thunderstorms in an afternoon on the continent than I have in my lifetime in Ireland. I also like mist. I love autumn because it's a fairly misty time of year around these parts. But on the whole, I'm not all that interested in the weather.

What I *am* interested in, however, is what *makes* the weather. Not in the sun and the rotation of the Earth and uneven heating, but in the physics and the laws and the variables. I love the maths here, and as I always say, maths is the language that science speaks. As I've progressed through my career over the past 27 years, I've also learned that I love to communicate all of this to the public. What makes weather work is something that many other people are interested in too.

So, my favourite part of the job is not and never has been standing in front of a blue screen pointing at where the clouds should be. My favourite part of the job is the science and the communication of it in the most effective way possible so that people can understand it – specifically, the sea area forecast.

What is the sea area forecast?

The sea area forecast is a part of the weather forecast, the primary focus of which is 'the protection of life and property on the island and our waters'.

These days the forecast output is sent directly from the computer models to the handheld devices of almost every adult member of the population. There are forecasts on social media, TV, print media, radio and the internet. It's in any format you can come up with and it's updated several times a day. There are weather enthusiasts taking every chart produced by models from around the world and commenting on them on internet chat rooms, sometimes fixating on snow and sometimes on high temperatures, and often of the opinion that they know better than the professionals. And sometimes they do.

But the sea area forecast is different. It is sacred ground. Although it's now available online as well as on the radio, it really hasn't changed since I started in Met Éireann, Ireland's national meteorological service, in 1995.

Starting at 0600, it's updated every six hours. Irish coast radio stations make a prior announcement of weather forecasts on Marine VHF Radio Channel 16 and then broadcast the forecast on the named relevant VHF Radio working channel.

It's broadcast by RTÉ Radio 1 just after 6 a.m., immediately after the news headlines, and again just before the pips for midnight. As a result, it has come to occupy an almost sacrosanct place in the day for many, its familiar, unique (and often incomprehensible) language acting as a wake-up alarm for an astonishingly high proportion of the population and sending another swathe of them to bed at the end of the day.

We, the weather forecasters at Met Éireann, work 24 hours a day, 7 days a week, 365 days a year. It's not the same person all the time – we have a roster and there are various jobs to be done. There are the TV broadcasts done from a sub-office within RTÉ. There's a shift that deals with daily requests from the commercial sphere, from racecourses to golf courses, pigeon fanciers to builders and movie makers. We have aviation forecasters who deal with the skies over the country and then we have the 'main desk' – the 'chief' or 'duty' forecaster. We've been switching and changing the name for as long as I've worked there and I don't think we've managed to agree on one yet. I'm partial to 'duty forecaster' myself.

The duty forecaster is responsible for updating the national forecast that goes out on the website and radio, they're responsible for the warnings issued on their shift, and they're responsible for the sea area forecast.

The job I do entails certain steps, and we'll talk about those steps in more detail as we go through the book. For now, let's just say that once you've put in the work you produce a picture of what you think the future is going to look like. There is a huge amount of uncertainty. Later we'll get to why there's even more uncertainty here on this island than there is in just about any other place on the planet, which means that sometimes it can go wrong.

I hate it when it goes wrong. And I do mean to use the word 'hate'. But I still understand that it *can* go wrong and I live with that. That's part of the job. An area of low pressure not caught correctly by the models, overestimating rain or underestimating cloud – these things happen and there's no point in getting bent out of shape about it. What we *can* do is apply due diligence in the processes of the job, meaning we've looked at every field, digested every variable, applied all our knowledge and experience, and produced the best product we can.

The sea area forecast takes six hours to prepare. There are six hours between updates, and for those six hours, the role of the forecaster is to assess the information available and update the forecast as necessary. Typing the forecast takes about 10 minutes, unless you're only using one finger and can't remember where the tab button is (or what it is for) – we've had those forecasters too.

I've heard it said that in order to build an apple pie from scratch you must first create the universe, and that's valid here. If I am to tell you how to make a sea area forecast I must first tell you how it all works. So in the next few pages, I'd like to tell you a little bit about why we forecast for the sea – which, as it turns out, is why we forecast at all. After all, the sea area forecast came first and it's still first now.

A LITTLE BIT OF HISTORY

The Royal Charter *Storm*

In August 1859 a passenger ship carrying around 450 passengers and crew set sail from Melbourne. Its final destination was the docks at Merseyside in Liverpool.

Australia was booming, and the gold rush that had started in 1851 was still drawing speculators from around the world. Energised young men, drawn by the promise of riches beyond their hopes in Europe, set off to make their fortune and take it back home. On this return journey, the ship was laden with gold and held a vast amount of wealth.

A relatively new type of ship, it was a massive 2719-ton iron-hulled steam clipper. Prior to the proliferation of steamships, global trade was limited to the times of year when it was possible to travel by sail. Doldrums, storms and monsoons meant the weather dictated when trade could take place.

Once steamships came into the picture, global trade could go on undisturbed all year round, with the steamers ploughing through wind and rain. The *Royal Charter* was a fast ship and was able to make the journey from Liverpool to Australia via Cape Horn in around 60 days. There were first-class cabins and plenty of room on board for passengers and crew to enjoy a relatively comfortable experience.

After a journey of more than 20,000km, with just 70 more to go to its home port, the ship met a storm just off the coast of Anglesey – the *Royal Charter* Storm.

On the night of 25 October, the on-board barometer started to drop dramatically as pressure fell with the advancing storm. As they passed the tip of Anglesey, they tried to pick up the pilot who would take the ship the final few kilometres to port, but the sea was too rough and the pilot was overpowered. Winds of storm force 10 on the Beaufort scale were recorded, soon rising through a violent storm to reach hurricane force 12.

It must have been terrifying for the men, women and children on board the ship, with their huge vessel tossed like a bean on waves that would have been crashing like thunder overhead as they huddled together in their staterooms and cabins.

Within sight of land, the ship was anchored and the crew cut their masts, but first one and then another anchor chain snapped, and the steam engines working at their capacity were no match for the gale force northeast wind. The ship was beached and then broken to pieces on the rocks with the rising tide.

With land so close by, many tried to swim, but in the turbulent seas and weighed down by their clothes and possessions, most didn't make it. Many were killed when they were bashed against the rocks and drowned. Joseph Rogers (a member of the crew, born Guzi Ruggier and originally from Malta) swam to shore with a rope and managed to rescue several people, earning him recognition from the townspeople of nearby Moelfre. Charles Dickens travelled from London to Moelfre to report on the tragedy that had claimed 459 souls. Twenty-one passengers and eighteen crew members were rescued or made it safely to shore. None of those saved were women or children.

The storm was named for the ship but went on to take more lives around the coasts of the UK that night and the next. In total more than 800 people lost their lives on the sea and land, with 133 ships sunk. In just one storm, this was twice as many as the total number who had died at sea around Britain and Ireland in the previous year.

There was also a financial loss. The ship was insured for more than £300,000, but it was estimated that the value of the gold lost was considerably more than that, in just one ship.

The first weather forecasts

Following the *Royal Charter* Storm, Vice-Admiral Robert FitzRoy began the first gale warning system. When a gale was expected, a network of warning cones would be hoisted at ports as a warning for those about to head to sea.

Vice-Admiral FitzRoy had been made head of what would later become the British Met Office in 1854. In the year following the 1859 storm, he distributed barometers, many of which he designed himself, to a network of 15 coastal stations around Britain and Ireland, including Valentia and Malin Head, which are still used today. He then had the readings sent to him in London, where he produced what he first termed a 'weather forecast'. These forecasts were published in *The Times* from 1861.

His warnings and forecasts went far beyond any that were available at the time and were at the forefront of the science of the day. Queen Victoria herself requested a personal forecast from Vice-Admiral FitzRoy for a crossing she was planning to make to the Isle of Wight. I bet he was up half the night worrying about how it went; just as I was when my friends asked me whether they should book a bouncy castle for their children's birthday party or when the film director and producer John Carney asked me if he needed a marquee for his wedding day!

Since then, the technology used to forecast gales and produce warnings has evolved, and the method of communicating the warnings has become more sophisticated, but the basic principle remains the same: we check the readings; we forecast a gale; we communicate it to the ships.

A gale warning is issued when winds are expected to reach gale force 8 or above. But there are many more hazards at sea besides the wind, and so the sea area forecast grew from the gale warnings. Now it provides an update on the wind, the weather and the visibility expected at sea.

The sea area forecast follows the same format at all times, designed so that even if all else fails and technology is lost, a ship at sea with a functioning radio should be able to use the information in a sea area forecast to navigate safely through or around bad weather to a safe port. It is this fixed format that has enabled it to occupy such a special place in our imaginations. Its rhythmic quality is both comforting and reassuring. The first part of the forecast tells if there are operational warnings in place, and in addition to the gale warning we issue what is called a 'small craft warning'.

While the gale warning covers the seas out to 30 miles off the coast of Ireland, the small craft warning extends to 10 miles off the coast and is designed for the smaller leisure or pleasure boats that are found nearer land.

Miles and knots

Although Met Éireann uses the metric system in fore-casting, nautical miles and knots are used at sea. A knot (kt) is one nautical mile per hour. The nautical mile is used to measure distance over the sea and is just a little longer than the miles we're used to on land – a nautical mile is measured as one minute of latitude and is approximately 1.15 miles. To quote the great Joe Pesci in the movie *Lethal Weapon* – 'Put them around boats and water and all of a sudden everything becomes nautical.' I'm paraphrasing of course – this is a family-friendly book and Leo Getz had a foul mouth.

The meteorological situation

Now we're getting to the guts of any forecast. Every day in the office the duty shift changes three times. At each changeover the duty forecaster hands over to their relief with a briefing. Every good briefing should begin with 'the meteorological situation', and every good forecaster should, having been given a good meteorological situation, be able to construct the rudimentary forecast for the next day, just like Vice-Admiral FitzRoy. That's basically what he was doing, after all.

While I don't want to get bogged down in the physics, there's hardly any point in going on if my reader – that's you – doesn't have the most basic grasp of this topic. As we move through the chapters of the book, I'll describe various aspects of meteorology in a little more detail, but for now, we'll manage with the basics.

Weather is defined by the movement of air, and air moves from areas of high pressure to areas of low pressure. In areas of *low pressure*, the air is rising and water vapour is condensing in the cooling atmosphere, forming clouds, which sometimes bring rain. In areas of *high pressure*, the air is descending, fizzling out the clouds and clearing off the skies. Areas of high pressure, also known as anticyclones, are associated with fair weather and light winds.

Areas of low pressure have lots more names – depressions, storms, cyclones, hurricanes and typhoons. Rising air circulates around a centre where the air is still – in a hurricane this is the eye and a hurricane, called a typhoon in the southern hemisphere, is a non-frontal area of low pressure, has a unique construction, and, of course, can be devastatingly destructive. We'll look more closely at storms and hurricanes later in the book.

A storm is a depression in which there are winds of force 10 or higher on the Beaufort scale, and a cyclone is the same thing as a depression. A tropical cyclone is a circulating low-pressure system that forms over warm ocean waters. (This is where some people start to mix things up. We don't mean to deliberately confuse people, but it does get a little complicated at times, and multiple names for the same or similar things don't help.)

When conditions are right, these develop further into hurricanes, with wind speeds of Beaufort force 12 or higher, and just to annoy people further, hurricanes are called typhoons over the Pacific Ocean. However, low pressure is low pressure, and whatever we call it the same thing is happening – in the northern hemisphere, air moves anticlockwise around the centre of the low, while in the southern hemisphere, this is reversed. And air correspondingly moves clockwise around an anticyclone.

Yes, I see the confusion. And no, I don't know why they called anticyclones 'anticyclones' when the air is moving clockwise and cyclones have the air moving anticlockwise. It's just another thing really smart people did to annoy the general public. They like to have their fun.

But really, there are simple rules that will help the budding meteorologist to navigate their way around

THE BEAUFORT SCALE ON LAND AND AT SEA

Beaufort	Mean wind (knots)	Wind descriptor	Sea descriptor	Land descriptor
0	<1	Calm	Calm (glassy)	Smoke rises vertically
1	1–3	Light air	Calm (rippled)	Direction shown by smoke but not by a wind vane
2	4–6	Light breeze	Smooth (wavelets)	Wind felt on face, leaves rustle, direction shown on wind vane
3	7–10	Gentle breeze	Slight (wave height 0.5–1.25m)	Leaves and small twigs in constant motion; light flags extended
4	11–16	Moderate breeze	Slight – Moderate (wave height 0.5–2.5m)	Raises dust and loose paper; small branches moved
5	17–21	Fresh breeze	Moderate (wave height 1.25–2.5m)	Small trees in leaf begin to sway; crested wavelets form on inland waters
6	22–27	Strong breeze	Rough (wave height 2.5–4.0m)	Large branches in motion; whistling heard in telegraph wires; umbrellas used with difficulty
7	28–33	Near gale	Rough – Very rough (wave height 2.5–6.0m)	Whole trees in motion; inconvenience felt when walking against the wind
8	34–40	Gale	Very rough – High (wave height 4.0–6.0m)	Twigs break off trees; progress generally impeded
9	41–47	Strong gale 3	High (wave height 6.0–9.0m)	Slight structural damage (chimney pots and slates removed)
10	48–55	Storm	Very high (wave height 9.0–14.0m)	Seldom experienced inland; trees uprooted; considerable structural damage
11	56–63	Violent storm	Very high (wave height 9.0–14.0m)	Very rarely experienced; accompanied by widespread damage

the weather, no pun intended. Buys Ballot's law is the easiest in my opinion: if you stand with your back to the wind in the northern hemisphere, then low pressure is on your left. Add this to the other rule above – that air moves from areas of high pressure to areas of low pressure – and we have the start of figuring out where the weather is coming from and going to.

So we start the sea area forecast stating exactly where the areas of high pressure and low pressure are around Ireland. This is the meteorological situation.

With depressions, we state the lowest atmospheric pressure and try to give as close an approximation to its location as possible – for example: 'A depression of 998hPa has its centre approximately 50 nautical miles northwest of Malin Head.'

Anticyclones are larger and slower moving, and we tend to state their highest atmospheric pressure with an indication of the geographic area that can best identify the location – for example: 'An anticyclone with a centre of 1024hPa has its centre over the North Sea.'

The scientific community understands a standard atmosphere to be 1013hPa (hectopascals). This is the number agreed to be the mean for the purposes of mathematical calculations, but high pressure and low pressure in meteorological terms are determined by the flow of air around the centre and not by the value in hectopascals. So we can see anticyclonic conditions with a centre of 1009hPa, for example, and we can have a depression with a central pressure of 1015hPa. It's the movement of air that's the defining characteristic, not the number.

Once a sailor at sea has the positions of the highs and lows, they should be able to start to figure out if they are going to face bad weather on their route. In brief, anticyclonic conditions generally mean fair weather, although there are caveats (there are always caveats). Anticyclonic conditions can also mean fog and slack or no wind, and these are no friend to anyone travelling by means of sail.

Now we move to the body of the forecast, and this is where we divide the area into blocks so that we can paint a picture of the weather as it is at the moment and how we expect it to change over the period of validity of the forecast. The forecast validity period is 24 hours, with an outlook for a further 24.

The headlands most used, going clockwise from the north, are Malin Head, Fair Head, Belfast Lough, Carlingford Lough, Howth Head, Wicklow Head, Carnsore Point, Hook Head, Roches Point, Mizen Head, Valentia, Loop Head, Slyne Head, Erris Head, Rossan Point and Bloody Foreland. Strangford Lough and Dungarvan are also identified but rarely appear, and how the forecast is divided is a matter of timing and the choice of the duty forecaster – some choose Dungarvan over Hook Head for reasons as simple as their having grown up in Dungarvan!

The headlands, sea area and forecasting

Recently, I've visited each of the headlands, and what follows in this book is my experience of my time there, along with weather and the meteorology that goes along with it.

When deciding which headlands to use in a forecast, we divide the area based on the wind and try to draw as simple a picture of the changing scene as possible. We are conscious that in an emergency the mariner could be relying on listening to the message over long-wave radio in possibly rough conditions and, more and more often these days, possibly not in their first language either. There is no benefit to trying to capture every small change in wind direction and strength over the period at the cost of making the forecast unintelligible to the listener. To quote my algorithm lecturer from university – who was quoting someone far smarter – keep it simple, stupid. KISS.

But back to the headlands. We go clockwise around them and start with the area that has the strongest wind. That's usually – but not always – in the west.

With depressions moving from the west to the east across the northern hemisphere, they have 'free passage' across the open Atlantic, skirting over the top of the Azores High (the anticyclone that is usually lying over the Azores) and making a direct line to Ireland. It is often at this point that the other blocking high-pressure area over continental Europe deflects the further passage

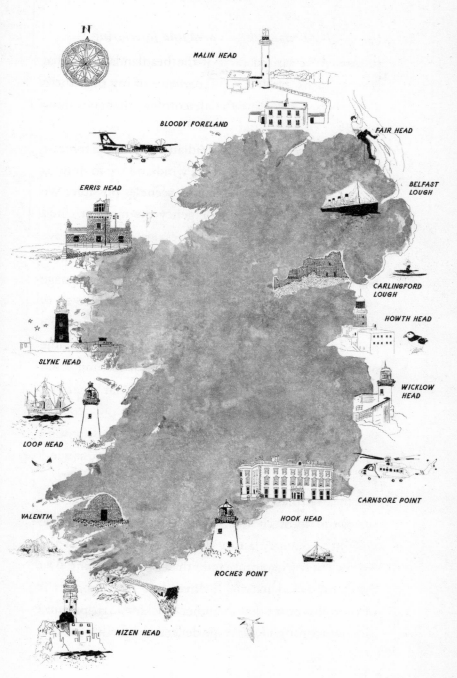

N

MALIN HEAD

BLOODY FORELAND

FAIR HEAD

ERRIS HEAD

BELFAST LOUGH

CARLINGFORD LOUGH

HOWTH HEAD

SLYNE HEAD

WICKLOW HEAD

LOOP HEAD

CARNSORE POINT

VALENTIA

HOOK HEAD

ROCHES POINT

MIZEN HEAD

east and the depression or storm moves northwards towards Norway, making for stronger winds on the west coast of Ireland than those experienced on the east. There is also the fact that as wind travels over land the effect of the friction of the land reduces the speed of the wind (and causes it to back somewhat). This means that the prevailing southwesterly winds along the coast of the west and south of Ireland are backed southerly or southeasterly and reduced along east-facing coasts.

We start with the strongest wind first, always conscious that a radio broadcast can drop at any point when out in turbulent waters, so it's best to have the worst of the weather accounted for earliest. Once we have the spatial taken care of, it's on to the temporal, because when forecasting the weather we have to consider variations over space *and* time.

It's only a 24-hour forecast, but a lot can happen in 24 hours on the waters around Ireland. To make it even clearer, the 24 hours are broken up further, and to avoid ambiguity, we've defined words that in other contexts can mean just about anything. For example, when I ask my son when he plans to get around to cutting the grass and he says 'soon', I can work out that the grass may well be cut any time between St Patrick's Day and Easter. When my publisher asks how soon they can expect the next chapter and I say it's 'imminent', they might expect it by either the end of the week or the end of the month

– depending on whether or not I'll have to hand back my advance.

For the purposes of the sea area forecast, if a change is 'imminent' it is expected to happen within the next six hours; if a change is expected 'soon' it is expected between 6 and 12 hours; and 'later' means any time after 12 hours and before 24. Sometimes when something is predicted to happen more than 12 hours ahead but we've already used a 'later', you'll hear things like 'by this time tomorrow' or 'by the end of the forecast period'. It wouldn't sound very lyrical or poetic to forecast something as going to occur 'later later'.

The wind is forecast on an eight-point compass: north, northeast, east, southeast, south, southwest, west and northwest. Within the body of the forecast we use the Beaufort scale, while in the outlook we use descriptive language – light, moderate, fresh, strong, gale force, storm force, violent storm and, thankfully only once in a very long while, hurricane force.

Although it's best to divide the sea area into as few groups as possible, two is most common, but sometimes there are three or as many as four. The Irish Sea can be either included in a group on its own or incorporated within an eastern group, and is often divided, commonly by terms such as 'the Irish Sea north of Anglesey'.

We keep the weather simple – it's not a case of some sunny spells and scattered showers at times in places. We stick with 'fair', 'fine', 'cloudy', 'rain', 'drizzle', 'mist' and 'fog', and include hazards such as 'snow' and 'thunderstorms' where necessary. Again, changes over time are indicated by 'imminent', 'soon' and 'later', and when the forecast might be complicated, sometimes added information is given by way of 'in sea areas north/south/east/west of' a relevant headland.

The language of the visibility section of the forecast is probably the easiest, although it's equally important – it's 'good' or 'moderate' or 'poor', often written as 'poor in precipitation, otherwise good'. This commonly happens when the weather within an area is varying over time – for example with a frontal passage.

With the main body of the sea area done, there are just a few more details to complete the picture. A warning of heavy swells is issued – usually and quite commonly on Atlantic coasts, where the fetch can be from as far away as the Gulf of Mexico. 'Fetch' refers to the distance travelled over the oceans from the one direction.

Moving on to the surface of the ocean, conditions here are determined by winds. But waves are complicated and, like sound waves, can be distorted when layered one on the other. If this book inspires you to travel to the headlands of Ireland, take some time to sit and watch the waves come in. You'll notice that periodically there'll be higher waves, and if you study the ocean in an almost meditative state, you'll become aware of the swell underneath the waves that ride over the top. When this swell rises above 4m, a warning is issued.

Descriptor	Wave height (metres)
Calm	0–0.1
Smooth (wavelets)	0.1–0.5
Slight	0.5–1.25
Moderate	1.25–2.5
Rough	2.5–4.0
Very rough	4.0–6.0
High	6.0–9.0
Very high	9.0–14.0
Phenomenal	>14.0

And then there are the ferry crossings – from Dublin, Rosslare and Cork to France and the UK. Again, it's simple enough language, and we use 'wavelets', 'slight', 'moderate', 'rough', 'very rough', 'high', 'very high' and 'phenomenal'. 'Phenomenal' is just that, waves of over 14m high that – when we view them from the safety of land – we can look at in awe and exclaim with all credibility, wow, that's phenomenal! I don't think I've used 'phenomenal' in the crossings, but I have seen it on my charts on more than one occasion on the west and southwest coasts. Some exceptional wave heights have been observed off our coasts over the years, and we'll cover these in the chapters to follow.

The final section of the sea area forecast comprises the coastal reports. This is probably many listeners' favourite part of the sea area forecast, curiously enough, particularly for those who are *not* mariners. This is also probably the most basic and intrinsic part of the forecast and dates all the way back to Vice-Admiral FitzRoy's initial crusade to protect seafarers around the coasts. The barometers he put in place in those 15 coastal stations around the country measured the observed atmospheric pressure, and the port-masters sending back the information to Vice-Admiral FitzRoy would include other basic observed facts such as wind direction and speed, the weather, the visibility and the changes observed in the atmospheric pressure – whether it was falling, rising or steady. When collected together, these coastal reports can be used to draw the crudest of barometric charts, and this is how Vice-Admiral FitzRoy was able to construct his forecasts.

It was a testament to the hidden popularity of the sea area forecast when one of the recurring familiar terms made it into popular culture in Glen Hansard and Markéta Irglová's hauntingly beautiful 'Falling Slowly'. The song won Best Song at the 2008 Academy Awards, taking our sea area forecast international – or at least a little bit of it.

The phrase 'falling slowly' appears in the coastal reports at the end of the sea area forecast. (The same coastal reports that were instigated at the behest of Admiral Fitzroy all those years ago.) This particular term refers to the pressure tendency, specifically, when pressure is falling at a rate between 0.5 and 1.9hPa over the course of three hours. Air pressure alone can only tell us so much. Pressure tendency is integral to producing a forecast. If pressure is high, then, in all likelihood, winds may be slack with possible clear skies. If we add in the tendency – that pressure is falling – then change is expected. The phrase 'falling slowly' tells us that the change is coming gently.

The reading of the sea area forecast, and in particular the coastal reports, provides a steady, familiar, almost lyrical break. The coastal reports have been successfully used as a sleeping aid by many insomniacs and are even used by hypnotherapists because of the steady, rhythmical pace of the words – no shocks, no surprises. Depending on the time of day that they are reading the forecast, different broadcasters sometimes use a different tone of voice when reading the list.

At 23.55, as the day ends for most listeners, the forecaster on the night shift at Met Éireann is four hours into the 12-hour shift. Conscious of the late hour and that many people use this late-night routine to send them off to sleep, they might read the last few lines of the sea area forecast – '... and finally, Belmullet, northwest, 23 knots, cloudy, 998, falling ... that's the forecast ... goodnight' – in a clear but soothing, gentle voice.

Just over six hours later, and 10 hours into the 12-hour shift, the same forecaster is back on air on RTÉ Radio 1 again. This time, although tired after a long night's work, they are conscious that just as many listeners have set their alarm clocks to wake them up at 6 a.m. with the early morning news followed by the sea area forecast. The tone of voice is now bright, awake, cheerful and sharp! It may be radio, but I swear to God, people can tell when you're smiling. So, now that we understand what the sea area forecast is, let's get things under way, people, and do it with a big smile on our faces.

In the following pages, I'll take you through both my journey of the past 27 years through weather science and the journey I recently made around our coasts, visiting the headlands of the sea area forecast. I'll guide you through the extreme weather that has passed over our sea area and the impact these storms and floods have had on the regions, and try to explain the science behind some of these phenomena. Each chapter will contain details about my visit to a headland and the people (and weather!) I met there, followed by (and sometimes interspersed with) the science behind weather phenomena or events particularly relevant to that area. Hopefully, once its mysterious language and patterns are explained, you will be left with a new-found appreciation and understanding of our island's unique daily sea area forecasts.

FROM LOOP HEAD TO SLYNE HEAD TO ERRIS HEAD

LOOP HEAD

52.558892, -9.931268

Loop Head is located at the westernmost point of
Co. Clare, at the very tip of the Loop Peninsula. It is
probably one of the most scenic and unspoiled areas in
the west of Ireland. The town of Kilkee is on the north
of the peninsula, its scenic horseshoe beach sheltered
from the Atlantic by rocks separating the land from
the turbulent and at times dangerous ocean and
forming safe swimming holes. The town is an ideal
tourist retreat.

It was late in September, approaching the autumn equinox, when the seasons turn and the length of day equals the length of night, when I decided on the spur of the moment to take a trip to Loop Head. This would be the starting point in my journey around the headlands of the sea area forecast.

There had been high pressure over the country for a few days, but insider information told me a weather front was expected. The big anticyclone that had been blocking out the Atlantic was going to break down, and a cold front would pass from west to east over the country. If I left Dublin early enough to avoid traffic, I'd drive through the rain and arrive on the west coast to sunshine and blue skies.

That was the plan, and it started out okay except for the traffic. No matter how early you set out, the M50 seems to never sleep, and as soon as I joined it, I saw a line of red lights stretching before me as far as I could see. Which wasn't very far – the rain was very heavy. Heavy and persistent, the weather front had a wave on it – this slows the eastward progress of the rain – but as I crossed the Shannon I noticed the first hints of blue in the sky.

Co. Clare looked very pretty in the sunshine, the border hedges trimmed and neat and gleaming from the recent rain. Everything still looked lush and rich despite it being the middle of autumn. Well, the middle of autumn according to the Irish calendar, of course, Meán Fómhair giving the clue (meán = middle, an

Fómhar = autumn, Meán Fómhair = September).
Meteorologically speaking, autumn only begins on 1
September, with the warmer months of June, July and
August getting the credit for summer.

Loop Head is a sparsely populated peninsula, and
Kilkee is a popular tourist destination on the northern
edge. So after driving straight through from Dublin, I
decided that this would be a good spot to recharge – both
myself and the car.

It was mid week and midday, and there weren't
many people around, but it was a glorious day. Reports
from back home were that it was lashing rain and that
it was going to rain all day there, but here a full day of
blue skies lay ahead. It was like I had planned it.

Looking for a bathroom, I wandered to the other
end of the car park and met a couple sitting in the sun.
The man was wearing a pair of swim shorts, and I asked
him if he intended to go in for a swim. He said he did,
but not just yet. Peter and Teresa Pane are from Limerick,
but have a holiday home here in Kilkee, and their inten-
tion was to sit in the sun for a while, then swim in the
bay. They'd been in Kilkee a week and would be staying
a while longer.

When I told Harm I was going to embark on this
tour of the headlands of my own country, he was very
sceptical. In the almost 20 years of our marriage, he
has become well used to the impulsive conversational
style of the Irish and our ability to start conversations

with strangers, but he's still a Dutchman. The idea that I would just go off and start talking to random strangers still seemed to him something a little short of odd.

But throughout our marriage, and with our three children, I'd found myself almost exclusively in the Netherlands for holidays because of family commitments, and I'd therefore neglected my own home turf dreadfully. I'd been hardly anywhere, and a tour around the headlands seemed like the absolute best thing to rectify the situation.

I've never had a shy moment in my life and have been talking to strangers since I learned to talk, which most people won't be surprised to hear was when I was very young. I'm a chatterbox and I find people new to me almost as interesting as science. So I knew that if I set off to go to a new place, I'd find people to talk to. I'm also a pretty lucky person, and while I've never won the lottery I tend to land on my feet, meeting really interesting people. So it was that I found myself sitting with my back to one of the oldest buildings in Kilkee, and Peter and Teresa set about telling me all there was to know about the town.

The boathouse was the original base for the coastguard, though they now have a newer building about 20m away. A big old rusting boat is up on blocks – the *Carmel Patricia* is blue and brown and has seen better days.

The current site for launching the coastguard lifeboats is marked off clearly, with warnings not to block the slipway, although it does indicate that launching boats is allowed, so long as cars are moved along quickly once the boats are in the water. This wasn't the original launch site, though. Peter pointed across the pretty horseshoe-shaped beach to another slipway that was the original spot for launching lifeboats. That site is much steeper and a little hairy in bad weather!

Teresa pointed out a giant iconic painting on the wall beside the original slipway – a painting of the revolutionary leader Che Guevara. It may be an odd enough site, but this painting is not mere graffiti.

On a sunny day in September 1961, a flight from Moscow to Cuba had stopped in Shannon. There, as a result of bad weather at the airport – probably fog! – the flight was grounded, and three of its passengers made their way west to Kilkee on the recommendation of the taxi driver they managed to pick up at the airport.

Jim Fitzpatrick, a local artist, was home from college and working in a bar in the Marine Hotel when three interesting looking men walked in wearing green raincoats with epaulettes. Jim was a follower of politics and was very interested in the revolution in Cuba at the time. He immediately recognised Che Guevara.

He struck up a conversation with the charismatic leader, and following his artist's instinct he drew a sketch of the man. The meeting inspired the stylised sketch you see today on students' bedroom walls, T-shirts and tote bags. Che Guevara and Jim chatted for a while, and Jim was fascinated to learn that Che was, in fact, partly Irish. His grandfather Patrick Lynch left Galway for Argentina in the 19th century. Of course he was Irish – sure, isn't everyone?

Peter talked to me about sea swimming, telling me that the beach at Kilkee is one of the safest swimming beaches along the west coast. There are yellow buoys marking out a line where speed on the sea is restricted to no more than 5kt, and further out from that line there are rocks. The sea on the day I visited was fairly choppy, and beyond the rocks I could see the spray rising over the Pollock Holes – popular swimming spots filled by the incoming ocean tides and warmed by the sun. This led to a conversation about a keen swimmer in the area, the world-renowned actor Richard Harris: A Man Called Horse to some, Dumbledore to others, but an unparalleled talent to all.

The Harris family had spent summers in Kilkee, and, having had to quit his beloved rugby as a result of tuberculosis, Richard turned to other sports, including sea swimming. He would swim the 2.5km across the bay, winning the annual Kilkee Bay Swim on a number of occasions.

Teresa pointed out the Harris family home on the other side of the bay, saying that if I could see the yellow house, then head out and up along the cliff, a few doors up I would see Richard's house with the plaque on the wall. Teresa and Peter told me that Richard was famous in the area for playing racquets, a cross between hand-ball and squash that was played against the sea walls surrounding the beach. He would hit balls against the wall almost all day, every day, and was a champion in the sport, winning the Tivoli Cup on four separate occasions.

This explains the statue of a young and handsome Richard in the car park of the restaurant at the base of the Cliff Walk. The statue is of a young boy with a racket drawn back to swing, facing out towards the Pollock Holes where the enigmatic young actor grew into a man.

I bade farewell to Peter and Teresa and set off for a run, heading around the picturesque beach and up the Cliff Walk, pausing, of course, to admire Richard Harris's house. At the base of the Cliff Walk I stopped to examine the swimming holes and chat with a local lady who'd been in for a swim. These were friendly folk, so my run was going to take a while.

There were a few people on the path up the hill, which starts with a steady rise and then peaks steeply near the top. It's tarmacadam but can get a bit tricky near the top in wet or windy weather – hence the handrail. The views were nothing short of spectacular looking both north and south. The aquamarine of the water as the waves crashed dazzled like emeralds.

I spent a few minutes there and headed back to the car, ready for the onward trip to Loop Head and the prearranged visit to the lighthouse there.

There's plenty of parking at the lighthouse, and despite it being mid-week there were a fair few cars already parked up and a couple of camper vans too. I suspected Americans and I was right.

As I made my way through the gates, I could see two men watching my progress with huge smiles on their faces. I was slightly distracted by the sight of the coffee van, with a few visitors taking cups of tea and little pastries away and sitting at the benches in the sunshine. I was looking forward to joining them after my visit to the lighthouse, but first I was looking forward to the lighthouse, which stood white and red against the clear blue sky, dazzling in the sunshine. There are a few houses there that were the lighthouse keepers' accommodation in the past, but the lighthouse dwarfs everything.

Up close, I could see the two gentlemen were wearing Commissioners of Irish Lights T-shirts. They were my tour guides, Martin and Simon.

The lighthouses around Ireland are operated by the Commissioners of Irish Lights – an organisation that is charged by the government with maintaining the maritime safety aids around Ireland's coast – and in a fairly recent tourist initiative you can now book online to visit or stay in one of the twelve 'Great Lighthouses of Ireland', one of which is the lighthouse at Loop Head.

Loop Head is the furthest point west in Co. Clare, and the lighthouse stands on the tip. There's been a lighthouse on the site since 1670, and this impressive building was completed in 1854, coincidentally the year of the foundation of what was to become the Met Office in the UK. Its light was mechanically operated until electrification in 1971, and the retired but operational mechanics are still in place, running up the centre of the lighthouse itself.

In the early days, the duties of lighthouse keepers who were operational in Loop Head included providing weather reports for transmission back to the Met Office. Reports were almost identical in form to what we have today – wind speed and direction, observed weather, visibility, barometric pressure and pressure tendency. These reports ceased when the last lighthouse keeper was withdrawn from service in 1991 following automation of the lights.

Despite no longer being part of the routine coastal reports, Loop Head is still very much part of the sea area forecast, its distinctive position regularly used as the point to divide the west coast. Looking south from the lighthouse, you can see Kerry across the mouth of the River Shannon, and northwards nothing but fields, cliffs and ocean.

In the grass on the edge of the peninsula is the ÉIRE 45 sign, made from rocks during World War II to alert aviators that they were above Ireland and had a few miles of flying yet to do to get to the UK. During the war years, the neutrality of Ireland was fairly selective:

British and US aviators who found themselves in trouble here were escorted to the border and given safe passage, while German aviators would find themselves interned for the duration of the war.

Leaving the lighthouse, I had a chat at the coffee shop and got myself a pastry and a cuppa to go. On advice from the lady who served me, I headed to the rocks just a few miles north. It's very easy to get lost in time around Loop Head. There are very few buildings that far out, and nothing at all spoils the landscape. On a sunny day the black rocks that characterise the landscape absorb the radiation of the sun and heat the air directly, making a warm blanket around a visitor who might just wish to sit and listen to the sounds of the ocean. One can easily imagine someone on a ship at sea, tossed about while trying to find their bearings. The lighthouse beacon flashes for 4 seconds in 20, so they know they're near Loop Head.

THE SPANISH ARMADA

Just a few miles north of Loop Head is Spanish Point, so named because of the Spaniards who were buried there after the wreck of the Armada ships. There was no sea area forecast and there wasn't even a lighthouse here to warn the Spanish Armada ships in 1588. Nor was there really an understanding of the volatile nature of the seas around Ireland, where the weather can change dramatically so quickly and where the swell at times extends from a distance of more than 6000km away.

Sent from Spain to defeat the British naval fleet by the Duke of Parma, the Spanish Armada was a fleet of over 130 Spanish ships, carrying almost 30,000 men, which had set sail for the English Channel.

A first attempt to set sail towards England about a month or two earlier had failed completely. From the start, this first fleet seemed destined for bad luck. After a month at sea, little progress had been made because of unfavourable winds. Badly packed food had already gone rotten and drinking water had gone stagnant. Sailing from Lisbon to the shelter of A Coruña in the northwest of Spain, they were met later by a fierce gale blowing in the Bay of Biscay. By the time the ships had made their way back to A Coruña and were ready to sail again with fresh provisions, it was July 1588. Morale was much higher as the fleet of over 130 vessels, galleons, galleys and supply ships hoisted sails for the English Channel.

The Duke of Parma and his huge army intended to cross from France to England with a view to wiping out the English fleet. In a sea battle on the coast of what's now Belgium, they suffered huge losses, while the English fleet came through unscathed.

The wind changed to south-southwest, and the commander of the Armada ordered that the fleet sail north around Scotland and wide around Ireland to limp back to Spain. But ongoing changes in the weather, so common for these latitudes, meant that the fleet was constantly trying to make corrections to its course. One report reflects the frustration of the navigators: 'We sailed without knowing whither through constant fogs, storms and squalls.'

Miscalculation of the Armada's position contributed greatly to its destruction. The navigators were also unaware of the effect of the northeastward-flowing Gulf Stream, which must have hindered the fleet's progress in sailing southwards along the Scottish and Irish coasts. The miscalculations and wrong estimates of their position meant that they ended up perilously close to the coasts of Scotland and Ireland.

By the time the fleet had navigated the North Sea and passed Scotland, provisions of food and water would have been seriously depleted, and as the Armada rounded the northern Irish coast, they came too close to the wild and unfamiliar coast of Ireland when they tried to restock the ships' provisions.

The prevailing westerly gales overwhelmed the ships and crashed them onto the rocky Atlantic coast of Ireland. The Spanish ships, built for the calmer waters of the Mediterranean, were lost. Many of these ships were spotted off the coast of Clare – four off Loop Head, two of which were wrecked, including the *San Esteban* at Doonbeg, with the loss of 264 sailors.

In all, along the 500km stretch of coastline from Antrim in the north to Kerry in the south, 24 ships were lost and between 5000 and 7000 men lost their lives, mostly by drowning. Those who survived the shipwrecks near the coast of Clare didn't fare better. By command of the English authority, all survivors were put to death by the sheriff of Clare, Boetius MacClancy. How differently would the history of the world have been written, had a sea area forecast been available?

We still have tragedies at sea nowadays, but they are undoubtedly less frequent. There have been advances in shipbuilding and dedicated lifesaving equipment in the first instance, and through education and the establishment of the met service and the sea area forecast there is much better information available to mariners.

The information we disseminate is also stored and becomes part of the historical records. When looking back on the events surrounding the destruction of the Spanish Armada, we rely on accounts that were written at the time and preserved by historians. But historians must work to unpick the facts of history from the remembered experiences of people, and that's where problems arise.

People remember weather and storms primarily in relation to how they themselves experience them. Undoubtedly, those who live in the exposed coastal areas of the Atlantic experience stormy weather more than those living inland or on the east coast. To be clinically accurate, when talking about the climate of Ireland

and the incidences of storms, we need access to the climate records.

Met Éireann's Valentia Observatory, for example, on average experiences gale force 8 winds 10 days per year. A steady strong wind, measured as a 10-minute wind (as opposed to gusts) above storm force 10, is relatively rare; and as of the time of writing, violent storm force 11 has only been recorded at the observatory on four occasions since 1940, the last occurrence on 12 February 2014, when Storm Darwin arrived.

Storm Darwin

Charles Darwin, an English naturalist, geologist and biologist, was born on 12 February 1809. He was the author of *On the Origin of Species* and the man behind the theory of natural selection. In modern popular culture, the Darwin Awards are awarded to people who place themselves in such hazardous situations as to naturally select themselves and their genes out of the human race, an example of which might be swimming off the pier at Salthill in Galway during storm-force winds.

Charles gets a mention in this book for two reasons. One, the ship that took him on his voyages of exploration, *The Beagle*, was captained by none other than the Vice-Admiral FitzRoy responsible for first using the term 'weather forecast' and for setting up the Met Office. Two, the storm we're talking about in this chapter is named after him.

Storm Darwin crashed through Ireland on 12 February 2014 (Darwin's birthday), after a particularly stormy winter season. The storm was named Cyclone Tini by the Free University of Berlin at a time before Met Éireann had started officially naming storms. With the winter season of 2013/2014 giving us Storm Jude and Storm Darwin, it also gave us the affirmation that names were going to be applied to storms, so it was best we take part in the process ourselves.

Storm Darwin started out in a slack area of low pressure south of Nova Scotia on 10 February. Moving eastwards across the Atlantic, it got caught up with cold air from the north and steered by a strong zonal Atlantic jet stream. (We'll talk about the jet stream again later in the book – its level of influence on Irish weather could merit a whole chapter of its own – but for now, understand it as an area of exceptionally strong winds that are found at approximately 30,000ft. The fact that this is the height that jet aeroplanes tend to fly at is not a coincidence; the planes take advantage of those strong tail winds, hence the name. And 'zonal' here indicates air flow from west to east along a latitude circle, without a significant north–south movement.) The jet stream rapidly deepened during 11 February, its central pressure dropping by 39hPa in a 24-hour period.

The definition of 'rapid cyclogenesis' at these latitudes is a pressure drop of 21hPa in 24 hours, and the drop in pressure observed by the storm that was later named Darwin was near to double this. A drop of 21hPa in 24 hours earns storms the term 'weather bomb', not something I'm particularly fond of – and I'll explain why when we get back to the idea later.

At the time, Storm Darwin was described by those directly impacted as the worst storm in living memory. Those looking at the data and perhaps less emotionally invested described it as a one in 20-year event. Certainly, it caused much damage. With a maximum wind gust on the day of 160km/h recorded at Shannon Airport, more than 215,000 homes lost power. There was significant coastal flooding and damage to buildings, and up to 7.5 million trees were destroyed, about 1 per cent of the national total.

SLYNE HEAD

53.398788, -10.233236

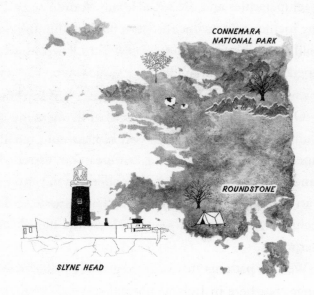

CONNEMARA NATIONAL PARK

ROUNDSTONE

SLYNE HEAD

Slyne Head is a rocky island outcrop off the furthest point west of Connemara in Co. Galway. One road and a couple of buildings, including a lighthouse, stand on the rocks. Slyne Head is 10km as the crow flies northwest of our coastal report station at Mace Head. Being so exposed on the west coast, Mace Head is often one of the stations to report on the strongest gusts from the storms that beat the west coast during the winter months.

My journey clockwise around the headlands brought me north of Loop Head to Slyne Head in Co. Galway. The winter I visited it had seen record-breaking low temperatures and the most severe winter storms in living memory across the northeast of the United States. A cold front moving through New York had caused a temperature drop from 6°C to –11°C in 9 minutes. Now that's what you call a cold front.

Over continental Europe, meanwhile, a plume of hot air rising out of Africa caused unheard-of January temperatures for at least eight European countries. In Poland, where a typical winter wonderland day would see an average temperature of just 1°C, it was 19°C. In Czechia it was almost 20°C when it should have been closer to 3°C.

Weather patterns are disrupted globally with climate change, but here in Ireland normal service was maintained during that time. The jet stream set up in a zonal flow directly onto Ireland sent one area of low pressure after another over the country, and this pattern brought cold weather and some light snow showers over hills and mountains and a little sleet for a while. Then, as my journey resumed, a little teaser of warmer weather brought rain and strong winds. As I set out for Connemara, the westerly wind was blowing a gale and streaming showers off the Atlantic.

It's a journey of over four hours by car from Dublin to Clifden, the closest large town to Slyne Head and my lunch stop on the way to my destination on that showery day. The Alcock & Brown Hotel dominates the landscape and is named for the winners of the *Daily Mail* prize in 1919 for being the first to cross the Atlantic Ocean in an aeroplane in less than 72 hours. John Alcock and Arthur Brown had flown a converted World War I bomber from Newfoundland, landing it in Clifden. (Later in the book there'll be more on aviation and the apparently endless lists of who did what first.)

Clifden is a large and busy town. Locally considered to be the capital of Connemara and a popular tourist destination on the Wild Atlantic Way, it has a lot to offer, especially food-wise. But as I arrived mid week and mid winter, I ate a simple lunch and headed on out the road.

The Atlantic coastline here is rugged and rocky, and the fields of Connemara are famous for their low stone walls, but the drive out to Slyne Head winds through a low-lying grassy landscape. The ocean seems to stretch all around as the road winds first around one headland, then another, and hilly islands just offshore break the seascape.

The showers constantly rolling in off the Atlantic on that day turned the sky a dark gunmetal silver, with rain spitting in a chilly downdraught, but they moved along quickly, and in minutes rainbows appeared in blue skies.

I could see the Connemara Golf Club taking up the space on the headland just before I reached a deserted camping site, and the road became narrower and narrower until I pulled in on a small gravel track.

There was a distinct lack of trees, but that was unsurprising considering the strength of the winds that lash these coasts from one end of the year to the next. How a young sapling could ever grow against the power of a barrage of gale force winds is clearly not a question that needs much consideration.

Buttoning up against the wind and icy showers, I spent about five minutes sitting in the car, bracing myself for the walk. I'm usually overly eager to walk, run or cycle, and am even known to brave the Atlantic, dipping my toes in the winter and my whole body in the summer, but sitting here looking out across the deserted landscape I found myself tempted to just cosy up with the car's air conditioning. After due consideration and having added a fleece-lined hat, fleece-lined mittens and my trusty 10-year-old Puma boots, I exited the car. It was cold and windy, and wind chill took several degrees off the real feel of the already frigid winter air.

AIR TEMPERATURE AND 'WIND CHILL'

The air temperature does not make an appearance on the sea area forecast. The mariner has little interest in the temperature, real feel or otherwise. The coastal reports make no mention of degrees Celsius, although the information is available – every weather observation includes temperature information, both the air temperature and the wet bulb temperature. (Combined, these two temperatures tell the forecaster the relative humidity of the air, and observed changes in the wet bulb temperature give us information about the position of weather fronts.)

Other national met services around the world routinely provide information in the weather forecast about the 'real feel' temperature, and bandy around phrases like 'wind chill factor'. In Ireland, not so much. There is a reason, and once again it comes back to our mild climate. Of course, we experience wind chill, but possibly not to the same extent as other places, mainly because, statistically speaking, our strong winds are generally southwesterly, coming off the Atlantic, and therefore aren't particularly cold. In fact, it's really never particularly cold in Ireland.

Wind chill or real feel relates to how people feel, and for this reason, there's no hard and fast scientific measurement. There are graphs, but the graphs vary depending on where you are in the world. The UK Met Office goes by a system called the Joint Action Group for Temp Indices, calculating wind chill based on how much heat is lost from a person's bare face at a walking speed of 5km per hour.

We feel the temperature of the air with our skin. When the temperature is high and we feel warm, we sweat to cool down. Moisture/sweat/water evaporates off our skin and we feel cooler, because converting this water (sweat) to a gas causes a net loss of energy at the surface of the skin. As work is done to convert the liquid to a gas, latent heat is required to break the chemical bonds of the liquid, and that heat energy is taken from the skin. When the relative humidity of the air is high, it's harder for moisture to evaporate into the air, and thus sweating provides little relief from the hot weather.

Conversely, when we need to keep warm we raise the hairs at the surface of the skin and trap that moisture at the surface, trying to prevent the moisture from escaping and to hold on to the heat. When the wind blows, it speeds up the rate of evaporation of the moisture off our skin and we feel colder.

When I visited Slyne Head that day, the wind was reaching gale force at sea, and an air temperature in and around 5 or 6°C made the real feel somewhere around freezing. My poor human face walking at about 5km per hour around the deserted headland couldn't take much more. I got back in my car and headed for the hotel, the wind reminding me of other storms that have hit this part of the country in the past and the difficult task weather forecasters have in predicting such storms.

PREDICTING STORMS: STORM ELEANOR

The winter season of 2017/2018 saw two major weather events that will be remembered for years to come. It started with Hurricane Ophelia in October and ended with Storm Emma bringing us blizzards at the end of February and the beginning of March.

These events were large-scale, major storms affecting almost all parts of the country, and they were well predicted in advance. Sometimes, however, forecasters can be (almost) caught by surprise by much smaller-scale storms that can be just as devastating, even though they affect a smaller area.

Halfway through the winter season of 2017/2018, in the early days of the new year, a small but intense storm developed just off the west coast. The computer models had signalled a few days before that a deep depression might develop just west of Ireland, but with a high degree of uncertainty regarding its track and intensity.

As it turned out, Storm Eleanor tracked right across the country from just a few kilometres north of Slyne Head to the east coast on the evening of 2 January. Wind speeds in the west exceeded those experienced during Storm Ophelia by quite some margin, and the highest gust of 156km per hour, reported at Knock Airport, was exactly the same as the highest gust reported during Ophelia (at Roches Point on the south coast).

Forecasting these smaller features in the weather remains difficult. Forecasters in Ireland are in the unfortunate position of being without land observations to our west. As storms like Eleanor move across the Atlantic Ocean, we have very limited upwind weather reports to work off. All we have are a few widely separated buoys at sea, an occasional ship sending back reports, or aeroplanes sending in observations.

On this occasion, the forecast model of maximum winds of the European Centre for Medium-Range Weather Forecasts (ECMWF) featured a narrow corridor of strong winds crossing Ireland and Northern Ireland.

Apart from the wind damage, there was a lot of coastal damage observed in Galway (where Slyne Head is) and Mayo (where the next headland I visited is) as a result of the combination of the storm-force winds and high tides. Parts of the city of Galway were submerged, and cars were seen floating in the streets. Waves of up to 20m were recorded by buoys at sea.

It wasn't over for Eleanor after it crossed from the west and left the east coast of Ireland either. The storm deepened further as it moved across the Irish Sea towards Scotland and Wales, causing similar storm and flood damage on the Welsh coast. It is not often that a storm causes huge disruption on both the Atlantic coast and the Irish Sea. Eleanor did, just like Ophelia had earlier that season. And despite the fact it was not a hurricane-size storm, the impact of Eleanor was almost as big.

ERRIS HEAD

54.307093, -9.996617

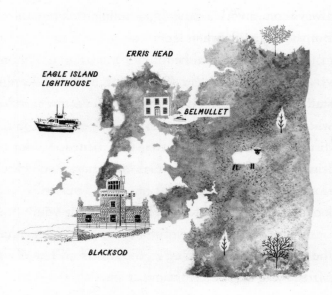

The northernmost point on the Mullet Peninsula in the west of Co. Mayo, Erris Head is a hike of just under 9km from the town of Belmullet, where there is a coastal report station. Erris Head is on the far end of the peninsula from Blacksod, the site of the famous Blacksod Storm.

The first thing to report on the visit to Belmullet and on out to Erris Head is that Belmullet is very, very far away. People have always reported Belmullet as being very far away, and for some reason that information always comes with a general agreement of 'yeah, yeah'. But the distances when looked at on the map belie the experience of the actual journey to this part of Co. Mayo.

The weather observation station at Belmullet was staffed up until 2012, so I've known several colleagues who worked in the region and they've all reported on the distances. I've known children who visited the nearby Irish college for summer courses and heard reports back from parents on just how far away it is. But again, one looks at the map and thinks, how far can it be?

It's possibly because of the scarcity of towns along the route. The measure of kilometres from Dublin to Mizen Head may be significantly more than from Dublin to Belmullet – but on the road it doesn't *feel* like it.

The Belmullet observation station is one of the oldest in the country, and the coastal report is the last listed on the sea area forecast: '... and finally Belmullet, West-Southwest, 25 knots, Gust 37 knots, Cloudy, 6 Miles, 1002, Steady.'

The station itself came into operation in 1956, its reports replacing those provided by the Sweeney family who operated the historic Blacksod Lighthouse, 16km to the southwest. (We'll return to the Sweeneys later.)

When I arrived in the town, low pressure was sending showers through on a strong and gusty southwest wind. It was cold and the wind wasn't making it feel any less so. Well wrapped up for the trip, I headed out to Erris Head for a look around and a bracing walk.

I drove out the road, past, well, not a lot of anything other than a handful of sheep and a couple of fields with cows still out. The number of houses diminished the further from the town I went, but didn't altogether disappear.

One thing that was noticeable in the area, though – as with the trip out to Slyne Head – was that there were no trees in the area. The fields I passed were a greyish green – and flat – with no trees surrounding them. As I headed further inland, the hardy hawthorns, their branches shaped by the wind, showed up in increasing numbers, but out on the headland that braces against the Atlantic winds, trees are unable withstand the wind.

When I attempted to get out of the car, I got a taste of why. It was a physical fight with the gale to push the car door open, and once out it was a struggle to stay on my feet. The sky was a steely grey, with only the thinnest glimpses of blue showing between the layers of clouds that rolled in non-stop from the west.

The ocean waves crashed along the coast, only about 100m away, but their sound was faint over the noise of the wind. The waves appeared huge, and the colours of the surf were purple and silver and green and blue and black. Van Gogh was in the wrong part of the world if it was colour he was looking for. One can only imagine the colours that would have appeared if even a little sunshine were to have broken through.

The town of Belmullet has a long association with weather observation, and the importance of this observation and associated coastal report cannot be understated. Belmullet, along with Valentia, is Europe's furthest point west, and until the relatively recent introduction of numerical weather prediction models, satellites and computer technology, Belmullet provided the first inkling that storms were approaching. As we shall soon see when we return to the Sweeneys, this was especially important on a particular June day in 1944.

WEATHER AND WAR

Weather bomb. Explosive cyclogenesis. Rain bomb. All terms I detest, as I've hinted already. I don't like words of war being used to describe weather phenomena. It's alarmist and unnecessary, and we've enough war to be dealing with in the world without making out that we're at war with the elements too. But then again, it's not new.

The association between meteorological terms and those of the industry of war dates all the way back to World War I, and here on the Mullet Peninsula the path of another war was shaped by a local resident. But first, let's go a little further back.

During World War I in Europe, the movements of troops through France, Belgium and Germany were slow, bogged down by sodden or frozen fields in the winters and often evenly matched armies for years on end. Trenches dug to hide young men resembled streets and roadways divided by minefields of no man's land, and these lines of trenches stretched for miles across the countryside.

Hundreds of miles away from the dirt and the blood, suited men circled tables in map rooms where the lines were drawn with pens on paper instead of barbed wire on trees and turned-over carts. Until early 1918, a blue line represented the location of the British or the Allied trenches, while the German trenches were overwritten with a red line. After 1918, the colours were usually switched to the other way around.

It was around this time that meteorology and weather forecasting became much more developed. It is often the case that war – more specifically the tools of war – drives scientific development. Accurate weather forecasts became more and more important with large-scale movements of men and machinery across continents and oceans and – at around this time – through the skies too.

It was a Norwegian scientist and meteorologist called Vilhelm Bjerknes who developed a theory of weather patterns identifying cyclonic development where cold and warm air collided. Associated with these cyclones, or areas of low pressure, were bands of cloud and rain, formed where warm air was forced to rise over colder air. These lines of cloud and rain were drawn on weather maps as 'fronts'. People had grown used to the idea of fronts from the war, and a red line depicted a warm front, while a blue line depicted a cold front. Arrows and half-circles added to the lines indicated the direction of movement of the front.

The Norwegian Model is still used today. It's simple, and the development of satellites and computer models in the latter half of the 20th century told us that there was a lot more going on out there, but the basic idea still holds its own as a useful way to easily describe the movement of air. Vilhelm Bjerknes also developed the Bergen School of meteorology and the basic numerical prediction equations that formed the basis of modern-day weather forecasting.

By the time of World War II, weather forecasting had developed further. When the Allies were trying to break the code to allow them to decipher what was being communicated by the Germans, they needed a starting point. They discovered that the Germans were collecting weather observations from their ships in the Atlantic and sending these weather reports back to base in coded communications using the Enigma machine. This code was changed daily and considered unbreakable.

Conditions such as air temperature and pressure, wind direction, visibility and wave height were sent back to German high command to become part of the weather picture. Each weather condition was represented by a different letter, but always in the same order. Using this information, cryptologists Alan Turing and Gordon Welchman were able to identify enough of the code to break it. So it was weather and maths that won the war in the end. Of course.

More practically speaking, weather reports, still in the same form that had been initiated almost a century before by Vice-Admiral FitzRoy, were needed from as many places as possible, and one of the most important of those places was the west of Ireland, and specifically Belmullet.

On 3 June 1944, a young woman on the Mullet Peninsula headed across the countryside to the post office where she worked. Then she and her family changed the course of history. Maureen Flavin Sweeney was working as a post office assistant, and part of that role involved sending the weather reports from the local weather station at Blacksod, where several members of the Sweeney family, notably the lighthouse keeper of the time, Ted Sweeney (husband of Maureen), worked in shifts gathering weather observations and reporting them to Dublin. In the early hours of 3 June, one of those readings indicated that pressure was falling rapidly, which happens when a storm approaches from the Atlantic.

What Ted and Maureen didn't know was that this information, from the most westerly station in Europe, was being sent to the headquarters of the Allied Expeditionary Force (in England), and would land on the desk of the US General Dwight D. Eisenhower, who was then Supreme Commander of the combined Allied forces.

After a phone call to Maureen from the meteorological office in Dunstable (England) to confirm the reading, the meteorologists in General Eisenhower's office were convinced to postpone Operation Overlord – what later became known as D-Day – for 24 hours. The Allies stood down all associated activities until the weather cleared. The break in the weather was also observed at Blacksod and reported by Maureen, and the invasion went ahead on 6 June.

Maureen only discovered more than a decade later what the influence of her report from 1944 had been. In 2021, the then 98-year-old Maureen was awarded a special honour by the US House of Representatives for her role in what may well have been one of the most weather-dependent events in the history of humankind – so far.

Maureen's story demonstrates the importance of weather reports and forecasting at an exceptional point in history, and reminds us more generally of the undoubted importance of the sea area forecast for the protection of life on the sea. But there are many more hazards at sea that make knowing your location critical to safety and

even survival. Today, satellites and radios guide navigators across the waters, and in their absence and before
their invention, celestial navigation was used to enable
sailors to pinpoint their exact location. That location is
identified by lines of latitude and longitude.

LONGITUDE, LATITUDE AND THE STORM
OF OCTOBER 1927

Long before it was a festival, longitude – and latitude
– involved lines drawn on the globe that were used to
navigate over land. Dating back to hundreds of years
BCE, the lines divided the Earth evenly, based on the
movement of the stars in the skies.

While measuring latitude at sea was straightforward, it was many centuries before another disaster at
sea prompted the British government to offer a prize to
anyone who was able to find a way to accurately measure
longitude. John Harrison's solution revolutionised travel
at sea when he presented his first designs in 1730.

The map of the world has changed in appearance over the years, with various empires considering
themselves the centre of the world through history. The
country in the centre the last time the maps were divvied
up was our near neighbour, the UK. And so the line of
zero longitude runs through Greenwich, in London.

Time is divided across the lines of longitude, and
Tonga is the first place in which a new year dawns every

1 January. Dividing the world evenly gives us the various time zones and gave Phileas Fogg an opportunity to win a prize by gaining a day. There are six time zones in the US and 11 span Russia.

Here in Ireland, for the sake of convenience, we have only one. We're in the same time zone as the UK and one hour behind the western countries of mainland Europe, such as France. As a result of the tilt of the Earth, summertime in the northern hemisphere manifests as 24 hours of sunshine over the Arctic, giving rise to the term 'Land of the Midnight Sun'. The furthest point west in Ireland is on the Dingle Peninsula, but because of the northerly location of Erris Head, it is there that the latest sunset in Europe is recorded.

At the summer solstice, usually around 20 or 21 June, sunset at Belmullet occurs at around 10:20 p.m., giving more than 17 hours of sunlight for the day, and with light still in the sky until after 11 p.m. Of course, in winter the nights close in early in this part of the world, and the days get very short for those trying to make a living through fishing on the coasts of Ireland.

If ever there was a reason to defend the need of the sea area forecast, one need only call on the October 1927 storm. On the night of 28 October 1927, a total of 45 lives were lost to the sea along the west coast of Ireland, with the storm continuing across the country and the Irish sea, breaching the sea wall at Fleetwood in Lancashire and causing death by drowning of another five people there.

In the early part of the 20th century, life on the islands off the west coast of Ireland was harsh. Communities relied heavily on fishing. Before the storm, the Mayo islands of Inishkea, north and south, supported about 350 people, mostly fully Irish speakers, but these islands were gradually abandoned following the tragedy, and today just two people live on the islands all year round, joined by another 15 or so during the summer months.

On the evening of 28 October, fishermen up and down the west coast of Ireland took advantage of a lull in the weather and set to sea in their currachs. Thirty currachs set out from the islands of Inishkea. Despite the falling barometer readings foretelling bad weather coming, the seas looked deceptively calm. Each currach carried two fishermen, and many of the fishermen were still in their teens. As the weather rapidly worsened and the sea conditions became increasingly worrying, 24 of the small boats turned back.

A currach is a small, traditional boat of wood and canvas still seen today in parts of the west of Ireland, and at the time it was the most popular fishing vessel among the small, poor communities of the islands. They were rowed between the islands and the mainland, bringing provisions for the islanders, and in their role as fishing vessels they provided food and commerce. At the time of the storm, there were also three trawlers recently purchased from the French, but with no harbour on the islands big enough to dock them, they were further up the coast at Blacksod for the winter. There was no lifeboat or lifesaving equipment on the islands of Inishkea at that time.

In the days leading up to the storm, weather reports from the archives show that a strong southwesterly airflow had been in place, with a succession of Atlantic depressions moving across and to the north of Ireland. This is a very common meteorological set-up at the area at this time of year. The polar jet would have been beginning to meander southwards by the end of October, where the heat generated by the movement of the oceans rising up from the tropics meets the colder sinking air from the Arctic, causing the strong jet stream to develop and then steer these depressions directly towards our island.

The observation station at Blacksod reported continuous winds of between Beaufort force 6 and 8 between 24 and 28 October 1927. Early on 28 October, strong southeasterly winds off the Mayo coast eased for a short while, before a strong northwesterly gale developed in the late afternoon. This would validate the witness accounts of deceptively calm seas early in the day as the centre of the storm moved over the area, encouraging the fishermen to take to the sea.

Historic synoptic charts of the time show the passage of a storm depression over the area with central pressure dropping by 18hPa between 27 and 28 October. A strong sea surge, the result of the long continuance of southwesterly winds across the Atlantic, contributed to the exceptionally treacherous sea conditions off the west of Ireland.

As well as the devastation felt on the islands of Inishkea, 26 fishermen drowned to the south in Cleggan Bay in Co. Galway, and another nine were lost from Inishbofin.

The news reverberated around the country and internationally as the impact of the horror began to be understood. The already impoverished communities were now facing into a winter with their families decimated and their livelihoods gone. Bodies continued to wash up on shore over the course of the next month – bodies that had been in the water for a month and which were unidentifiable except by their clothing. Relief flooded in from abroad, with money coming from Australia and America. The sum of £36,000 was amassed and then distributed amongst the islanders, but with their young men gone, it was little condolence.

The population of the islands of Inishkea never recovered, and the battle to hold on to a life on the remote islands off Blacksod was lost.

FROM BLOODY FORELAND TO MALIN HEAD TO FAIR HEAD

BLOODY FORELAND

55.156969, −8.282719

BLOODY
FORELAND

DUNFANAGHY

GLENVEAGH NATIONAL PARK

Bloody Foreland is a sparsely populated area in the northwest of Co. Donegal. You can fly to Donegal Airport, voted one of the most scenic airports in the world – it's right on one of many beautiful beaches in Donegal. A 30-minute drive will bring you to an area so named as to have visitors expecting a scene of a great battle, but the area is actually named for the red hue of the rocks at sunset.

Anticyclones are large and slow-moving or – often-times – stationary. Depressions or areas of low pressure tend to be much more mobile and are often over a smaller geographic area. As an area of low pressure moves eastwards across the north Atlantic and over Ireland, what tends to happen is that the anticyclone usually lying over the Azores (the Azores High, mentioned earlier) will extend a ridge of high pressure over Ireland. You'll often hear this mentioned in the meteorological section of the sea area forecast as a 'temporary ridge of high pressure'.

As I set off to visit Bloody Foreland, the jet stream was continuing to direct low-pressure systems on a direct path for Ireland, and on this day, a temporary ridge had built. This brought great good fortune: under a ridge, the sky clears. There was a little bit of sunshine and not a huge amount of wind. Given that I was visiting one of the windiest spots in the country, this was indeed fortunate.

I stayed nearby in Gweedore – or rather Gaoth Dobhair, as this is an Irish-speaking area. Gaoth Dobhair boasts the largest Irish-speaking parish in Ireland, with more than 4000 Gaeilgeoirí. The place name Gaoth Dobhair means 'gentle wind', and if the wind is experienced as gentle in the northwest of Ireland, it must be due to the shelter of the mountain ranges that span the region. Because I arrived at night, it was the next day before I noticed the dominating outline of Mount Errigal looming over the landscape.

Mount Errigal looks like a young child's rendition of a mountain. Its 751m peak appears to form a perfect right angle if turned on its side. On my visit, a bobble of cloud obscured the top of the highest mountain in the Derryveagh Mountains and the highest peak in Donegal, and the rocky face gave a greyish Alpine appearance to the sky.

It's a short drive to the coastal area of Bloody Foreland, but the roads are narrow and winding, and more than a little nerve-racking, with steep declines right at the road edge, tall cliffs on the other side, and nowhere to go should you meet an oncoming car. Thankfully, there weren't an awful lot of people around to encounter head-on.

I saw a small community of houses – and it looked like a year-round community as opposed to the mostly holiday homes that appear at so many picturesque coastal regions of Ireland – right down at the coast that Google Maps described as Bloody Foreland beach, although 'beach' is a stretch of the imagination. Bloody Foreland itself is a region more than just a beach or a headland. It was hard to concentrate on the road, as the views all around were breathtaking. Every angle southeast to northeast opened up more cliffs, more coves, more inlets.

I stopped and took advantage of the break in the weather for a short walk in the sunshine. It was the depths of winter, so the landscape was brown and grey. As in Mayo, there was a scarcity of trees, but there was an abundance of fern, and apparently in the autumn its changing foliage turns the hills red, adding to the red caused by the setting sun on the rocks which, as mentioned earlier, contributes to the foreboding name of the area.

Heading out of Bloody Foreland and the west of Donegal towards Inishowen and my next stop – Malin Head – I drove through Glenveagh National Park, feeling like I was on the set of an alien movie. The road was good, newly resurfaced and fairly straight and wide, but I dreaded the thought of an unlikely breakdown. There was no evidence of human activity for miles in any direction, just a long, clear, empty path with the mountains brown and grey all around – intimidating and a little bit awe-inspiring too. Since I was travelling alone, I took extreme care.

Road conditions reminded me of a song that can be heard playing on the radio around Christmas time – one of those tragic American country songs – and a line in the song goes something like 'the roads are getting worse as the freezing rain turns to snow'. But the scientific reality – not that science is ever in question in an American country music song – is that when freezing rain turns to snow, the road conditions are actually likely to be improving.

The exceptionally mild autumn prior to my visit to Donegal had given way with a sudden shift as mean daily temperatures (that's the temperature of the day as a whole, over the day and night) went from two to three degrees above the seasonal average to as low as five degrees below in the space of a week. The sudden cold snap came as an anticyclone from Siberia made its way towards Greenland and sent a cold easterly then northerly Arctic wind over the country.

Warnings followed for snow, sleet and hail showers, and icy conditions and – a fairly rare one for Ireland – freezing rain. And this brings us conveniently to another science topic of relevance to weather – the various states of water.

WEATHER AND THE CHANGING STATES OF WATER

I once heard a radio presenter refer to 'liquid water' and then laugh, thinking he needed to correct himself, as if water didn't occur as anything other than liquid. But, of course, it does. In much the same way as when everyone calls a vacuum cleaner a hoover and everyone else knows what they're talking about, it's generally understood that water (H_2O) is a liquid.

And yes, water is usually (perversely, not always) a liquid between 0°C and 100°C. But it changes state at both 0°C and 100°C. Changing the state of water from a solid to a liquid requires a huge amount of energy, and the reverse releases a huge amount of energy. Changing the state of water from a liquid to a gas also requires a huge amount of energy – and takes an age. A watched kettle, anyone?

So the radio presenter was validly referring to liquid water. There's liquid water (or just 'water', as we refer to it); solid water (or 'ice', as it's more commonly known); and gas (what scientists call 'water vapour' but the general public more commonly call 'steam').

Water is possibly the primary reason behind everything about science being so complicated. It can also exist as a liquid, gas and solid all at the same time in the same place, constantly changing state and releasing and absorbing energy.

In the case of frozen precipitation, mostly it's easy. The air gets frigid and the cold extends through the atmosphere, the water vapour condensing to cloud. Then precipitation forms little ice crystals that bunch together – yes, every single one of them different – to fall as snowflakes to the ground.

When it's cold but not quite cold enough, liquid waterdrops and ice crystals fall together, and we get a mix of rain and snow that we call sleet. There's also hail, which is formed in cumulonimbus clouds, and there's the somewhere in-between phenomenon called 'soft hail', or 'graupel' (from the late 19th-century German word Graupel) as the people of some more northern parts of Europe, who experience it much more, refer to it.

When snow falls, we can see it. It's white! And crunchy. We know it's there and take care when walking, cycling or driving on it – well, most of us do. When it's very cold, usually at night in these parts of the world, and dew forms and settles on the ground and surfaces and then freezes, it forms frost and is also white. And bumpy. And we can see it and take care when walking on it or cycling on it or, for most of us, driving on it. Hoar frost is also bumpy, and it forms when water vapour freezes and attaches itself to surfaces. Granted, frost can get very slippy too, especially when it has been walked on, refrozen and then walked on again, but freshly landed, it's mostly navigable.

And then there's freezing rain. If the freezing rain were to turn to snow, conditions would be improving. Freezing rain is a very rare occurrence in Ireland, simply because it rarely gets cold enough. It is a more common hazard for our neighbours in the UK, as temperatures there in midwinter can get lower than here because of the strong influence of the weather systems of continental Europe. Once again, our warming prevailing southwest winds keep us cosy.

But cold alone is not enough – there's an added aspect that makes freezing rain occur. Precipitation falls as ice crystals through a sub-zero layer of the atmosphere. Then a layer of warmer air above the surface, usually due to an approaching warm front, melts the crystals to their liquid state. These little drops of liquid are barely above freezing and may still be in a liquid state at 0°C as they continue to fall towards the surface. Through the warm layer of air, now the layer of air at the surface is once again below freezing, cooling these water droplets further. But the droplets stay in their liquid form. In order to freeze to a solid, they need to encounter condensation nuclei, which are microscopic particles suspended in the air. Without sufficient time, they stay liquid. We refer to these liquid drops of water as being 'supercooled'.

At the surface, the ground is frozen – at or below 0°C. And when the supercooled droplet hits the surface, it instantaneously spreads out and freezes. The tiny drops join up and form the smoothest, thinnest, clearest layer, and it's invisible.

Black ice.

A lethal hazard. We can't see it. It doesn't shine or glitter or sparkle. It doesn't give off a crunch when we carefully put one foot in front of the other. We just slip. And fall. Or slide in our cars.

Thankfully, on that winter day I made it through Glenveagh National Park without mishap and headed towards my next destination – one of the windiest spots on the island.

MALIN HEAD

55.381781, -7.372358

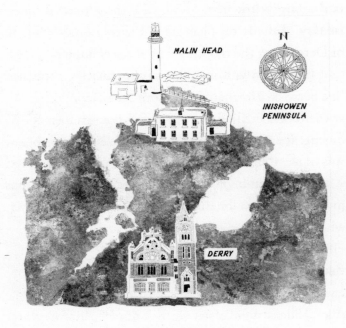

Malin Head on the Inishown Peninsula of Co.
Donegal holds the title of the furthest point north on
the island of Ireland, further north than anywhere
in Northern Ireland. It's the site of one of the original
coastal stations, and remains a prominent name in
the sea area forecast today, as it divides the country
so neatly east to west. It is also the first of the coastal
reports to be read out: 'And now for the coastal reports,
starting with Malin Head ...'

A visit to the northernmost point of Ireland takes you to the Inishowen Peninsula. At the narrowing of the peninsula where it meets the mainland, Letterkenny is the largest town in Donegal and is located at the estuary of the River Swilly. On the other side, the city of Derry is on the mouth of the River Foyle.

Inishowen juts out from Ulster, dividing the coast-line between Northern Ireland and the Republic. Right at the top, the ÉIRE 80 sign was written into the land during World War II, notifying aircraft that this land was in the Republic.

I had been invited to stop for lunch at Farren's Bar in Slievebawn, a small community nearest the head, where Ali Farren, the local promoter of all things Malin Head and the proprietor of the nearby caravan park, had volunteered as guide for the afternoon, or rather I should say he *was volunteered*. Martin McDermott, a local coun-cillor, had got wind of my plans to visit the headland and had given Ali the heads-up. When I entered the pub I was greeted by Ali with a 'so you're the lady that does the weather visiting today'. I was quiet and he was dubious, but I was soon up at the bar and we were chatting like we'd been friends for ever.

The bar is the furthest north on the island, and that midweek afternoon in early January, its clientele were mostly Americans, propped up at the bar as new locals or sitting eating lunch, eavesdropping on where to go to find the most stunning views this place has to offer.

Pauline provided us with sustenance in the form of a very tasty soup and sandwich, with a little sweet thrown in with the obligatory cup of tea for after, and the visit was so comfortable and friendly that it was easy to become too relaxed and risk missing the views altogether – that far north it gets dark early! At the winter solstice, just before Christmas, the sun sets there at 4 p.m. in the afternoon; by early January it has gained an extra 18 minutes.

It was a lovely day and there was blue in the sky. We set off in Ali's car and he gave me the history of the region – ancient and more recent – as we took the scenic route towards Malin Head. Ali was born and raised in Farren's Bar, and his grandfather worked as a weather observer at Malin Head before it was incorporated into the newly formed Irish Meteorological Service in 1936.

As I drove north, Ali pointed out all the houses that had been occupied by previous workers at the station. The weather station at Malin Head was officially closed on 31 December 2009 and its operations replaced by an automatic weather station at the same location on 1 January 2010.

Malin Head (Cionn Mhálanna) has a long history of communication with ships. A lookout tower, known locally as 'The Tower', stands in a commanding situation at the point known as Banba's Crown. In Irish mythology, Banbha (or Banba), daughter of Ernmas of the Tuatha Dé Danann, is the patron goddess of Ireland.

Communications and the location of the observation station at Malin Head have been vital for shipping. The coastline around Malin Head has some of the most treacherous waters in the world, with many hundreds of shipwrecks recorded. There are more ocean liners and German U-boats sunk off this stretch than anywhere else in the world, and most of them were casualties of the world wars in the first half of the 20th century. Malin is reported to be the windiest station in Ireland, and storm force 10 winds are recorded at the station most years.

The current buildings were constructed in 1955, when Irish Meteorological Service personnel commenced regular hourly weather observations. However, the station records go back to 1885. Personnel employed by Lloyd's Insurance Company, who staffed and operated a local coastal signal station, were the first to prepare and transmit regular weather reports. Coastguard officers continued this work until the early 1920s, when Ali's forebears took over, passing the job through the family's hands until Met Éireann was established.

Ali reports that his grandfather knew the weather on 'the Head' so well that he could write his weather reports in advance, then head back into the family bar for a few hours. I cover my ears and sing *la la la la*, but Ali says his grandfather is long dead and punishment for this crime is no longer a concern.

On my visit, we stopped at a few different viewing spots around the area, and the views were, needless to report again, stunning. At one such stop Ali pointed out a little rock off the coast that is officially Malin Head, but it's not the furthest point north, and not where the ÉIRE 80 sign is, nor where the signal towers are, though they're not far away.

I headed over there and donned my hat. It was still a beautiful day and the winds had died down from earlier in the week, but there was still a cold wind blowing from the northwest, and hats were needed. Ali pointed out another slightly larger island further out to sea. This island had been up for discussion earlier on by the American at the bar – the rock from the island is reportedly millions of years older than that of the mainland, suggesting the movement of plates and a fault line nearby.

Inishtrahull is about 10km off the headland, making it the northernmost island of Ireland. The island houses a lighthouse and was inhabited until 1929, but the lighthouse was automated in 1987. It has recently been reported that the deer that had lived on the island alone for the past 20 years had died. It had been thought he had died with the rest of the herd many years ago, but he had been spotted by birdwatchers on the island, with locals marvelling that the big old buck had lasted alone out there all those years.

Then there were more tourists, struggling to get a selfie with the ÉIRE 80 sign in the background, so we offered to help out. I didn't catch the accent, but they were from somewhere in Europe, I guess. They were friendly and we shared our comments on the landscape. Ali was clearly proud to be the representative of Malin Head.

The one thing that is missing at Malin Head is any sort of visitor centre or tourist facility beyond the wiggly Wild Atlantic Way sign and a few printed information posts. There's so much to see here and the site has a lot of history, including early telegraph and ship-to-shore communication technology. Ali told me that 300,000 tourists had visited the headland the previous year, but that the number could be a lot higher. At the time of writing, he's trying to encourage an electric charge point for the community centre, as he'd like to see the development of the tourism industry in the area.

As the northernmost point on the island of Ireland, the place has a lot to offer. The views and beaches are a huge draw in the summer, when the area can enjoy very pleasant sunny days. The coastal showers that plague the western coastal counties tend to form a little further inland, so the peninsula itself manages to enjoy a slightly drier climate.

In wintertime, meanwhile, the area around Malin Head is a great spot to catch the Northern Lights. The aurora borealis lights up the night sky with eerie green hues that dance over the horizon as solar winds interact with the magnetic fields over the polar regions. The opportunity to see the lights is not so easy to come by, their elusiveness making the chance to catch a glimpse of them all the more appealing. The conditions need to align: the solar winds; the atmospheric conditions in the ionosphere – about 80km above the surface of the Earth; and the weather conditions closer to the surface in the area between Europe and the Arctic. Skies need to be clear – not a condition that's reported regularly in the middle of winter at the station at the most northern point on the island of Ireland at Malin Head.

In September 1961, the skies were far from clear when the remnants of Hurricane Debbie passed close to here. The storm affected much of the country, with wind records broken in the west and north. Malin Head recorded what, at the time of writing, is still the highest windspeed at that location – 98kt (that's 122km/h) – the highest wind speed recorded in September on the island of Ireland to date. On the same day, a 10-minute mean wind of hurricane force 12 was recorded.

Debbie took an unusual path after it formed. At first, there was nothing strange about its track. A tropical depression was born off the coast of Senegal in early September. It then started to move westwards across the Atlantic, like most of these developing tropical systems. After reaching the Cape Verde Islands as a full tropical storm, where it caused a plane crash that killed 60 people, it went on to move westwards, further away from any land.

For days, nothing was seen of Debbie until a Dutch KLM plane encountered the storm. From the data taken by the plane, it could now be confirmed that the storm had grown into a full hurricane. Its path had changed as well, now taking a more northerly course instead of a westerly one.

Further north the path changed again as it got caught up in the southwesterly flow. This is the direction our weather comes from. Although Debbie had lost its hurricane status, the storm gained momentum as it accelerated on its way to Ireland, lifting on the strong jet stream. The jet also deepened the storm again, and by the time Debbie reached Ireland on 16 September, winds off the west and north coasts had reached hurricane force again. I've already done a fair bit of name-dropping so far in this book, and perhaps now, as we leave Debbie behind, it's about time we delved behind the scenes of all this storm-naming.

NAMING STORMS

The area close to where Debbie passed experiences more named storms than any other part of the country. Hurricanes may hardly ever happen in Hertford, Hereford and Hampshire, but if they did, we might want to know the answer to the question – what category were they?

Hurricane Sandy in 2012 peaked at category 3 but was actually a category 1 hurricane when it caused most of its destruction up along the east coast of the US.

Ophelia was a category 3 hurricane just a day before it moved northwards to make landfall on the Irish south-west coast.

Katrina was a category 4 hurricane when it made landfall a few miles southeast of New Orleans in August 2005, claiming 1800 lives and causing widespread devastation.

Hurricane Lorenzo was a category 5 while it lingered around the Cape Verde Islands before meandering north-wards to sit relatively harmlessly off the west coast of Ireland for a few days in 2019.

The 'category' we refer to here is the Saffir–Simpson Hurricane Wind Scale. Developed in 1971, it takes the name of civil engineer Herbert Saffir and Robert Simpson, a meteorologist who at the time of its develop-ment was the director of the National Hurricane Center in the US.

Although most deaths directly caused by hurricanes are the result of floods, mainly associated with sea surges as ocean waters are driven on to land, the scale relates solely to wind strengths.

The table opposite is taken from the National Hurricane Center's website.

One might be forgiven for thinking we live in the centre of the world, and that's just human nature

CATEGORY	SUSTAINED WINDS	TYPES OF DAMAGE DUE TO HURRICANE WINDS
1	74–95mph 64–82kt 119–153km/h	**Very dangerous winds will produce some damage:** Well-constructed frame homes could have damage to roof, shingles, vinyl siding and gutters. Large branches of trees will snap and shallowly rooted trees may be toppled. Extensive damage to power lines and poles likely will result in power outages that could last a few to several days.
2	96–110mph 83–95kt 154–177 km/h	**Extremely dangerous winds will cause extensive damage:** Well-constructed frame homes could sustain major roof and siding damage. Many shallowly rooted trees will be snapped or uprooted and block numerous roads. Near-total power loss is expected with outages that could last from several days to weeks.
3 (major)	111–129mph 96–112kt 178–208km/h	**Devastating damage will occur:** Well-built framed homes may incur major damage or removal of roof decking and gable ends. Many trees will be snapped or uprooted, blocking numerous roads. Electricity and water will be unavailable for several days to weeks after the storm passes.
4 (major)	130–156mph 113–136kt 209–251km/h	**Catastrophic damage will occur:** Well-built framed homes can sustain severe damage with loss of most of the roof structure and/or some exterior walls. Most trees will be snapped or uprooted and power poles downed. Fallen trees and power poles will isolate residential areas. Power outages will last weeks to possibly months. Most of the area will be uninhabitable for weeks or months.
5 (major)	157mph or higher 137kt or higher 252km/h or higher	**Catastrophic damage will occur:** A high percentage of framed homes will be destroyed, with total roof failure and wall collapse. Fallen trees and power poles will isolate residential areas. Power outages will last for weeks to possibly months. Most of the area will be uninhabitable for weeks or months.

– apparently people in other parts of the world feel the same way. But although we hear mostly about the Atlantic storm season here in Europe, there are tropical cyclones developing into what we call hurricanes in the Pacific and Indian Oceans. They just call them typhoons or cyclones. Just to confuse us, as I mentioned before. They're all the same thing.

The National Hurricane Center in the US is based in Florida. It was founded in its current form in 1955, the year after the particularly rough hurricane season of 1954, when three hurricanes – Carol, Edna and Hazel – affected the densely populated areas of the Mid-Atlantic states and New England, and got the attention of the politicians in Washington. The official naming of hurricanes from an official list goes back to 1953. And, of course, storms were named before then too.

In fact, storm-naming has gone on from as far back as people were paying attention, with storms often taking a name from a saint, for example, if they fell on a named saint's day, or the name of a particularly famous ship that was sunk (remember the *Royal Charter* Storm?) or an island that was devastated.

When weather forecast offices started to track storms and communicate their developments and expected tracks, they often identified the storms by their position according to latitude and longitude. It became clear very quickly that this was subject to confusion. There was often more than one storm active at any one time, and which storm was going north and dying and which was going west and strengthening was getting mixed up in the translation of the message.

For a short time, there was a plan to name storms by the phonetic alphabet, and a list was at the ready, but then someone changed the phonetic alphabet and that plan was shelved.

There had already been a practice of naming storms using women's names in Australia at the end of the 19th century, and during World War II the army and navy adopted the practice when plotting the movements of storms in the Pacific Ocean.

So between 1953 and 1978, the National Hurricane Center worked off a list of women's names, with men's names being added from 1978. Since then, the alternating male/female list of names has been compiled from an agreed list in cooperation with the World Meteorological Office. When all the names are used up, we move to the Greek alphabet, and when a storm is particularly devastating, the name is retired for all time. The lists are laid out for a long time in advance.

Naming a storm doesn't assign a personality. It doesn't assign a gender. A hurricane is a physical phenomenon and has no gender. The name does, however, serve a very important purpose. *It aids in the effective communication of a clear message that is crucial in times of emergency.* The same purpose applies here in Ireland and Europe, and the development of winter storm-naming collaborative groups grew from the same need to make clear and effective communication a top priority of weather services.

In October 2013, a rapidly deepening and fast-moving low-pressure system interacted with an ex-tropical storm and was steered by a strong Atlantic jet stream. It had all the ingredients necessary for the making of a very serious weather event, and it was. The storm centre moved across the south of Ireland, through the south Irish Sea, to bring devastating winds through England on St Jude's Day.

The name 'Storm Jude' was attributed to several sources. One suggestion that a clerk at the British Met Office named the storm was denied, but whoever started the name, Twitter took it over, and soon the UK main-stream media carried the name into the public realm.

At that time there were no official storm-naming procedures in place in Britain or Ireland. In Germany, the Free University of Berlin's meteorological institute has made it possible for quite some time for members of the public to 'buy' the name of a depression or an anticyclone. The public can volunteer any name and then pay for the privilege of that name appearing on weather charts. Anticyclones are more popular; they stay on the charts longer. In fact, the Free University of Berlin named Storm Jude 'Christian'.

The Swedish Meteorological and Hydrological Institute named the storm 'Simone' based on their name day list; the European Windstorm Centre gave the storm the name 'Carmen'; and the Danish – who got the highest wind speed from the storm, by the way – named it 'the October Storm 2013', a bit more of a mouthful and perhaps why they afterwards renamed it 'Allan'.

Following the disastrous naming of Jude/Christian/Simone/Allan, it was clearer than ever that some sort of system should be put in place so that the value of improving the efficiency of communication by having a name for a storm could be achieved.

In time for the winter storm season of 2015, the meteorological services of Ireland and the UK collaborated to introduce the public to the process of storm-naming. The same storms that go through Ireland often affect the UK, although not always. The public are invited to submit names, which are then assessed by a joint team comprising members of both the UK and Irish services. In a style like that of the National Hurricane Center's list, the list is alphabetical, alternating between male and female names.

The first storm named from the list was Abigail in November 2015, and in time for the winter season of 2019 the Dutch joined the same group. There are also collaborations between the other met services of Europe, based on how storms affect different regions. France, Spain, Portugal, Belgium and Luxembourg are together; in the eastern Mediterranean it's Greece, Israel and Cyprus; and in the central Mediterranean it's Italy, Slovenia, Croatia, Montenegro, North Macedonia and Malta.

Naming storms aids in the effective communication to the public at times of risk to life and property from severe weather. From time to time a storm at sea will bring force 10 winds – forecast and observed – within the 30 nautical miles of the coast of Ireland that the sea area forecast covers; but without occasioning the naming of the storm. A storm is only named when the impact is expected on land, unlike the naming of hurricanes, which are always named in order to make tracking the development and expected path of the storm straightforward for all concerned.

FAIR HEAD

55.220803, –6.153914

Fair Head takes us into Northern Ireland, but the weather and the sea area forecast know no political boundaries, and we continue uninterrupted around the island. Fair Head is another scenic spot, and a hike of about 8km out of Ballycastle will get you there in under two hours. It's only about 20km to the Mull of Kintyre, and it's easy to see Scotland from the headland on a clear day. You could probably swim it if you were feeling energetic.

A nticyclonic conditions over Europe blocked the passage of frontal depressions right through the autumn and into the early winter that preceded my visit to Fair Head. The consequence was a very wet autumn over Ireland as weather fronts stalled along the western edge of the country. It also made for one of the mildest Novembers on record, with a record-breaking highest low temperature at Shannon Airport. The mild and humid weather was possibly more noticeable overnight, as temperatures remained unchanged from day to night.

When temperature is shown on a graph, it waves up and down in a pattern similar to a sine wave, cresting in the day and troughing in the night. We call this changing pattern 'diurnal variation', and it happens, naturally enough, because the sun – the source of heat – shines during the day and radiates heat into the ground, warming it up. The Earth then radiates it straight back off again, warming the air directly in contact with the ground by conduction. The warm air then rises, and the heat is transferred through convection.

At night, when the source of heat is gone, the heat is radiated away from the ground. When it's 'gone', the air cools, and in the winter, when the air temperature goes below freezing, we get frost as water vapour in the air freezes into tiny ice crystals to give surfaces a pretty white covering.

This is the normal ebb and flow of heat. Heat also moves around the Earth 'en masse', so to speak, carried by the movement of the air masses. This is called 'advection'. The warmer and colder air moves by a process of advection.

During that wet autumn, the warmer air was adrected in by large anticyclonic conditions over Europe, and there was little sign of diurnal variation. The advection process overrode the normal input and output of radiation.

The day I set off to visit Fair Head, the air temperature was about 18 degrees, even though it was mid-November. The depression out to the west of Ireland was deep and the anticyclone to the east was large, and the resulting squeeze on the isobars put me on the road with a strong southerly wind at my back. There was something of a southeasterly aspect too, so waves crashed all along the east coast as I wound my way north.

I initially planned to visit Carlingford Lough on this day, but it was such a nice day that I stayed on the road north and went all the way to Ballycastle, a few miles to the west of the northernmost headland in Northern Ireland.

Rain from a warm front had cleared early in the morning, so we were in a warm sector when I set off for a hike out to the headland. There were a few spots of drizzle and it was breezy – typical conditions for a warm sector in Ireland. The 'warm sector' is the space between the progress of a warm front and the associated following cold front.

Fair Head stands an imposing 200m high and dominates the landscape. A very popular spot for climbers, the sheer cliff face, as high as 100m in places, boasts one of the biggest areas of perfect climbing rock in all of northwest Europe. I did not intend to do any climbing, however. I had no interest in strapping myself to a rope and hanging off a sheer wall, but I took a walk out along the coast all the same.

I was quickly warmed as I walked, despite the strong breeze at my back. The high temperatures had me shedding layers along the route, and I was glad I had left the hats and scarves in the car.

The area is hilly, but there's a good walking path along the coast, with a golf course split in two by the narrow roadway. The town falls away slowly, and the number of houses diminished as I moved closer to the headland. The coast is rugged around here and the scarcity of buildings was clearly not a hindrance when it came to choosing sites for filming the very popular *Game of Thrones* – Fair Head is the location of the Cliffs of Dragonstone. I could see Scotland, and it wasn't even that clear a day.

Fair Head makes a regular appearance on the sea area forecast, along with Malin Head. The east–west split divides the strong winds that affect the west of the country from the lighter winds experienced further east, and with Fair Head taking such an elevated position, it's an obvious choice when dividing up the areas.

As I made the trip back to the hotel, I noticed that the clouds in the distance were starting to lower, and soon the cold front was catching up with the only recently passed warm front. The air had been warm and humid, and so this cold front, with much cooler, clearer Atlantic air behind it, was going to be sharp. Sure enough, with only about 1km to go to the safety of the hotel, the heavens opened and I was soaked in minutes.

I dried off as best I could and sat down to enjoy a cup of tea and some cake, watching from my vantage point as the clearance came quickly in over the town to the east, rainbows forming in the sky over Rathlin before the sun shone on Fair Head itself and the black rock glistened golden in the low-angled sun.

The setting – large hotel windows, a fire burning cosily, and hushed waiting-staff slipping barely noticeably across the thick carpet to bring more tea – couldn't have been more comfortable for storm-watching. Unlike the Indiana Jones-like storm chasers in the film *Twister* – where meteorologists Helen Hunt and her ex-husband Bill Paxton chase down tornadoes across Oklahoma with a vengeance made personal by tragedy – I like to do my storm-watching from a place of safety – high ground, thick walls, warm fire.

At times like these I remind myself of the fortunate position the people of this country enjoy in the world, especially in these modern times, when, as seldom as it may happen compared with other parts of the world, very severe weather can be forecast well in advance and some preparation can be made for the protection of life and property from the elements. In 1839, many of the residents of Ireland had no such luxury.

The 'Night of the Big Wind'

When Lloyd George's government introduced the old age pension to the UK and Ireland in 1908, payable from 1 January 1909 and with many clauses for eligibility attached, recipients were required to be 70 years of age or older. Records of births, deaths and marriages in Ireland, particularly in rural Ireland, were scrappy and unreliable in the early to mid 19th century, and many people didn't have birth certificates.

The old age pension was distributed through the post offices of the UK and Ireland. It was not a lot of money – 5 shillings for a single recipient, 7 for a couple when the husband was over 70. It was kept deliberately low so that the population would be encouraged to make their own money into retirement and prepare for their own pensions, but it was very gratefully received by the poorer of society, some people bringing flowers or apples in gratitude to the post office where it was distributed. Winston Churchill, in conversation with Lloyd George, said, 'It is not much, unless you have not got it.' To verify whether someone met the age criteria in Ireland, some were asked to account for their whereabouts on the 'Night of the Big Wind'.

Beginning on the afternoon of 6 January 1839, the significant windstorm that swept through Ireland and the UK caused widespread devastation. Hundreds died and thousands were left homeless. It is estimated that a quarter of the homes in the greater Dublin area were destroyed. With a barometric pressure of 918hPa recorded, it was the worst storm in at least 300 years, according to remembered history recorded during that time.

That new year had started cold. There was snow on the ground. Snow is rare enough in Ireland, with most of us able to recall less than a handful of times in our lives when we were able to make 'proper' snowmen here, but there had been a period of heavy snow on the night of 5 January. Then a warm front went through and put the country in a warm sector for a short period of time. In early January this meant cloud cover and mainly dry conditions, and rising temperatures melted the snow quickly.

Then came the associated cold front. A deep area of low pressure in the mid Atlantic, the driver of the frontal systems, started to move towards Ireland. Winds are strongest just ahead of the cold front, and the first reports of stormy weather came from the coast of Co. Mayo around noon on 6 January.

By midnight the winds had reached hurricane force 12 on the Beaufort scale. That's a sustained wind speed of greater than 64kt. But it's a very difficult thing to interpret strength of wind speeds from numbers like knots and kilometres per hour. In fact, the difficulty in interpreting the strength of winds by individuals, and removing the individual's perspective, bias and subjectivity, is central to the Beaufort scale itself. (See the scale on page 14).

The scale is named for Admiral Francis Beaufort, an Irishman and sailor. Francis Beaufort was born in Navan, Co. Meath, on 27 May 1774, to parents descended from French Protestant Huguenots. His father, Daniel Augustus Beaufort, was a map-maker, a Protestant clergyman and a member of the Royal Irish Academy. His mother was an heiress of William Waller, of the historic Allenstown House, demolished in 1930 by the Irish Land Commission. Francis was the second son and had three sisters. He grew up in Wales and Ireland until his teenage years, when he left school and took to the seas. He was shipwrecked at the age of just 15, as a result of an inaccurate chart. He consequently developed a keen interest in the need for accurate charts on the sea, and his most significant accomplishments are centred on nautical charting. He continued his education informally for the rest of his life and became a well-respected scientist and mathematician.

As a sailor, Beaufort rose through the ranks during the Napoleonic wars and spent all his leisure time studying latitude and longitude, coastlines and astronomy, with the goal of constructing charts. In 1829, at the age of 55 – an age when most of his contemporaries would be considering retiring to a nice cottage and living out their later years in rest – Beaufort was appointed the British Admiralty Hydrographer of the Navy, where he served 26 years, longer than any other hydrographer.

The wind strength scale that bears his name evolved over the course of more than a century from its first official use in the early 1800s. The credit was given to Beaufort because it was he who first officially set down a standard for how wind speeds were described at sea. Until then, while observations were taken on the sea by mariners, one person's gale could have been another's stiff breeze.

In January 1839, housing for the common people across rural Ireland was rudimentary – usually single-storey dwellings with thatched roofs. Many of these were destroyed, and indeed many of the larger structures, such as the landlords' country estate houses, lost their roofs too. The storm swept across Connacht, Ulster and Leinster, with structural damage evidenced across the entire region. Between a fifth and a quarter of all the homes in Dublin were damaged in some way, many destroyed entirely. Newly built churches, barracks and factories all reported significant damage, and at sea 42 ships were destroyed while trying to ride out the storm at sea. Most of these were on the west coast.

The surge associated with the hurricane force winds that swept the ocean ashore along the west coast caused severe flooding. Many of the hundreds of deaths that resulted from the storm were from drowning.

In the aftermath of the storm, crops that weren't destroyed by the floods were ruined by the wind. Hay and feed that had been set aside for livestock to last through the winter were swept away on the night's wind. So the devastation went on long after the winds had eased and the waters had receded. Animals were starving, and starving peasants had to try to rebuild their homes while still in the depths of a freezing winter.

To put yourself in the position of the people of rural Ireland that night is near impossible. It was 6 January in Christian Ireland. Nollaig na mBan held significant religious meaning, and this strange weather that started with drifts of snow menaced the families as they sought shelter in the darkness.

The wind moved swiftly through the Beaufort scale of wind, picking up rapidly right across the northern half of the country as the steering depression at sea sent the lashing rain and floodwaters across the landscape of Ireland.

It was only a couple of weeks past the winter solstice, and darkness would have set in early that day. With the approaching weather front there would have been no moon or stars to give any light, and no candle or fire would have withstood the gales. It would have been complete darkness. When roofs were lifted off structures that should have given shelter and walls started to fall, families would have had to gather in ditches, but these ditches rapidly filled with water from surges, rain and the recently melted snow.

It's very hard to appreciate the terror these people must have felt that night. And for those who lived through it, trying to prove their eligibility for the pension nearly 70 years later, they'd certainly have remembered where they were on the 'Night of the Big Wind'.

FROM BELFAST LOUGH TO CARLINGFORD LOUGH TO HOWTH HEAD

BELFAST LOUGH

54.690801, –5.785043

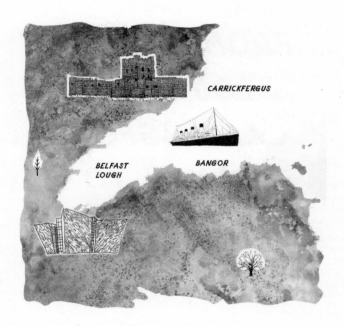

Belfast, the capital city of Northern Ireland, straddles
the counties of Down to the south and Antrim to the
north, and Belfast Lough lies in between. The city
grew on the linen trade and was once the linen capital
of the world, though this may not be as well known
today as another great export that sailed out of the
lough. The Harland & Wolff cranes are a landmark
in the city that will always be known as the birthplace
of the Titanic.

Belfast Lough marks a boundary at the northeastern edge of the island, where winds from the north can undergo strengthening as they are funnelled between the headlands of Scotland and the northeast coast of Ireland just a few miles away. The northern boundary between the Irish Sea and the North Channel can see turbulent waters.

The River Lagan flows into the estuary that forms the border between Down and Antrim, and the lough's sea-side boundary is the line between Orlock Point and Blackhead. The lough is long, wide and deep, and without the inconvenience of strong tides, which makes it a favourite for sailors.

The coastline north and south hosts many other towns and villages, with Bangor on the southern coast and Carrickfergus on the northern shore. The Belfast coastguard actually operates out of the marina at Bangor.

If any of the points around the coast of the sea area forecast shouts 'maritime', it must be Belfast, with its history of shipbuilding and of building possibly the most famous ship in the world.

Visiting Belfast is a whole-family affair. There are museums, galleries, colleges, shopping and, of course, the Titanic Museum. With the improved motorway, a trip to Belfast by car takes just over 1.5 hours from our home on the northside of Dublin.

Heading directly to the docks, now referred to as Titanic Quarter, we could not but stand in awe of the giant Harland & Wolff cranes. The museum built to accommodate the thousands of international visitors to the site of the building of the RMS *Titanic* was designed to bring to mind the scale and expanse of the famous ship.

Standing on the viewing platform on the first floor and looking into the vast depth of the quay, the visitor gets the impression of how dominating the ship would have been in the area. After the tour, I felt the heart-breaking realisation of the devastation that was the loss. The people of Belfast were rightly very proud of the incredible vessel. She was the product of years of their hard labour and quality workmanship. Blood, sweat, tears and lives were given to her construction. That she was lost, along with 1496 souls, was a personal grief for the city to mourn in shock.

Even though the tragedy occurred more than a hundred years ago, it left an enduring and almost visceral scar, and the cause of the accident that sunk the 'unsinkable' liner is still up for discussion. But we do know that an iceberg dead ahead ripped a hole through 6 of her 16 compartments, opening up the ship to the freezing Atlantic.

The north Atlantic Ocean is unsurvivable for long at any time of the year, but in April it is almost at its coldest – the water doesn't warm up until the summer months, reaching its peak around August.

Ice from the glaciers at Greenland break off periodically and float into the ocean. The ice forming on the glaciers is fresh water, so the icebergs float in the sea. There's a difference in the density of each and, obeying Archimedes' law of buoyancy, about 90 per cent of the iceberg is below the surface.

On 14 April 1912, four days into her maiden voyage, the *Titanic* received six separate warnings of sea ice. But it was the dead of night, and by the time the watchmen on the lookout spied the dull figure breaking the blankness of the dark horizon, it was too late. Travelling at 22kt, the ship tried to turn but was too ungainly. Disaster struck.

In recent years, ice in the oceans has become a threat for other reasons, and there is much discussion of the melting ice caps at both the North and South Poles, and the extent of sea ice and the consequent impacts of the disaster currently facing humanity – that of climate change.

There are many implications to consider. Melting sea ice is fresh water and has a different density from salt water; and large amounts of fresh water spilling into the oceans would upset the global oceanic circulations. The global circulation of warmer water currents is a vital part of the distribution of the heat held in the planet, and any change in this will clearly affect the climate patterns of the world.

The melting of the ice fields and snow over land also releases into the atmosphere vast amounts of greenhouse gases stored in the frozen ground beneath, exacerbating the problem of warming that we already face. In addition, the ice caps reflect the radiation of the sun away from the surface of the Earth, reducing the amount of heat that comes into the planet and the atmosphere. This is due to the 'albedo' of ice.

Albedo is a measure of how much light – or radiation – a surface reflects. The various coverings of the Earth have different albedos. The Earth is covered principally by water, snow/ice, forests or vegetation, and deserts. Each of these surfaces reflects the radiation of the sun to a lesser or greater extent. Fresh snow reflects almost all of the radiation that lands on it, whereas water or wet soil absorbs almost all of it.

As the ice caps shrink in size, the amount of the sun's radiation that is reflected away from the atmosphere is reduced, and this contributes to an increase in warming. As with all things related to the weather and climate, there are many factors to consider, with many implications to feed into the equations. But as computer science develops and machines become much more powerful, scientists are closing in on a consensus as to what the future will look like if the Earth continues to heat at current levels.

Over the years, that increased computer power has greatly changed the way weather forecasts are produced. Computer models have replaced hand-drawn charts in most weather offices (though in Met Éireann, until very recently, charts with isobars and weather fronts were drawn every hour with a soft pencil). Nevertheless, computer models have been the main tool for short- and long-term forecasts for a while now. Improved and more frequent radar and satellite images have also made forecasting more accurate, especially in the short term.

The way we consume the weather forecast has also changed a lot over the years. Radio and television have always been the main sources for people to get the latest weather forecast. Newspapers also publish forecasts in text form, though with the delay between printing a paper and distributing it, the forecast would always be a bit dated.

PRECISION, ACCURACY AND YOUR SMARTPHONE WEATHER APP

With the advancement in technology and the improvements of computer-calculated weather forecasts with the aid of supercomputers, the road has opened to delivering automatically generated weather forecasts for every location possible on Earth. This started more or less as a one-symbol forecast for major cities around the world in your daily newspaper or on TV stations like CNN.

Now, with a smartphone in everybody's pocket, people also have at hand the hour-to-hour weather forecast for wherever they are. These forecasts can look up to a week ahead or sometimes even longer! And this is very convenient, of course. At any time, you may ask yourself: will I get a shower if I hang my washing out tomorrow? Or will it rain during my match on Saturday? The app on your phone will give you a very accurate answer. But the question is: is it really accurate? Here lies the problem.

It might not be accurate, because the movement of air or flow in the atmosphere is by nature unpredictable, even beyond a matter of days at times. On a global scale, there are so many movements of weather systems with different speeds and directions that it is impossible to fit each of these exactly in a computer model. There are countless small, unpredictable atmospheric fluctuations, which mean that a computer forecast never starts out 100 per cent right, because we simply cannot account for all these random movements.

If you don't know where you are right now, how can you predict where you will be in the future? If we don't know what the atmosphere is doing right this second, how can we accurately predict what it will be doing in the future? We can't, of course, but we can make a pretty good approximation. So what we do is we make a pretty good approximation of the 'where we're at now' – and we call these the 'starting conditions'.

Small changes in the starting conditions can have larger implications further along the timeline. The further from the start we move along a timeline, the larger the implications of any errors in the starting conditions. This is the basis of chaos theory.

After about 10 days, the forecast often shows little or no accuracy in predicting the behaviour of the real atmosphere. There is not much we can do about this. There will always be a limit to how far and and how accurately we can forecast the weather, no matter how much the computer models develop.

This brings us back to the forecast for your location on your smartphone. You can see now why that shower that is forecast for 2 p.m. on Saturday, maybe a week away from now, might not be completely spot on. It is a mix-up of precision and accuracy. The forecast is very precise, but it is not necessarily very accurate.

And that's why forecasts on television and radio will always have value. The TV forecast may say something along the lines of sunny spells and passing showers, while your phone might say it will stay dry for the day. That is just because the computer model didn't calculate a shower for your exact location. You know now from the written or verbal forecast that you can still expect a few showers.

By contrast, the sea area forecast is for 24 hours with an outlook for a further 24. There is a very good accuracy associated with weather over this short period of time and for this relatively small geographic area, which makes it vitally important for those who will be affected by the weather at sea.

CARLINGFORD LOUGH

54.049784, -6.178025

Another few miles along the coast and we come to
Carlingford Lough. On the border between the Republic
of Ireland and Northern Ireland, Carlingford is a
popular tourist town nestled between the Mourne
Mountains in Co. Down to the North and the Cooley
Mountains in Co. Louth to the south. The shelter of the
mountains provides a microclimate that gives the lough
its popularity for water sports and the associated tourism.

It was all very last minute when I finally made the decision to head to Carlingford Lough. In a kind of 'the cobbler's kids are last shod' style, I was always going to be typically last minute getting to the headlands closest to where I actually live, because I always had the attitude of 'sure I can just pop up the road'. And sure enough, Carlingford is just 'up the road' these days as the M1 motorway speeds out of Dublin and takes us almost all the way to this charming little harbour town.

Carlingford Lough appears on the sea area forecast dividing the often lighter winds of the Irish Sea and the stronger winds that can sweep around the northern coast. The lough is a physical manifestation of the border with Northern Ireland, with the small historic town of Carlingford on the southern border and the larger town of Warrenpoint on the Northern Ireland side.

It was cold as I set off. The air was clear and fresh and the skies were blue, but as I drove along the M1 I could see the large, towering cumulonimbus clouds that were crowding on the sea. The showers were staying off the coast and it promised to be a beautiful day on land. The air was stable to the air temperatures over land and unstable to the now much warmer air over the sea. With the sea surface temperatures on the Irish Sea still at more than 14°C, the air over land was only 9 or 10°C. Those few degrees might not look like much, but they make the difference between the air rising (quickly) through the atmosphere to form cumulonimbus clouds and there being no clouds at all in the sky.

A small boat setting out to sea would need to be mindful of the conditions around these large clouds, some overshooting the tropopause at the top of the troposphere to create a large anvil-shaped cloud made up of ice crystals. (The troposphere is the layer of atmosphere closest to the Earth's surface; the next layer up, coming before the stratosphere, is the tropopause.) These conditions can generate thunderstorms with down-draughts. A downdraught is a cold wind that shoots down from the cloud and out at the surface, and this gustiness can come on suddenly and cause dangerous conditions for those caught unawares. On my trip to Carlingford, I saw a beautiful example of one of these clouds – it looked exactly like a profile of Yoda sitting on top of the Mourne Mountains against a clear blue sky.

Carlingford was originally a walled town and is littered with the remains of its medieval buildings. King John's Castle is the first to make an impression, taking pride of place high above the harbour, and you can visit during the summer with guided tours.

The lough used to come right up to the town wall, but land was reclaimed over the centuries and a large train station was built to carry goods and people to Newry and beyond. The tourist office now sits in the building that was the station, and you can follow the tracks northwards to the next town along – Omeath – on a scenic, low-lying, flat walk.

I climbed the stairs to the castle and walked across the bridge, and finding myself on higher ground I then walked back into the town and through another arch, this one on the building marked as the gaol. The arch marked the place where the native Irish would pay to gain access to the walled town in order to sell their goods to the Anglo-Normans living inside.

I stopped at the Village Hotel for a delicious brunch before taking a hike along the higher ground. The way up to the walkway is a steep incline, and I was a bit puffed by the time I turned onto the rough path that winds along the treeline on the mountain that looks down on the town and the lough below.

It's a really extraordinary place. The lough itself is a hub of activity, with busy harbours and ferries going to and fro. The surface of the lough was calm on this day, and it is a sheltered area generally, ideal for the water activities it has become quite popular for. There are also indoor and outdoor adventure and activity centres all around the town.

The views along the walk, which, after the relatively steep climb up, was flat and easy, are stunning. The mountains and sea make for a painter's paradise, with lots of details from the medieval history of the area preserved. From this distance they looked as though they were only built yesterday, and as though an Anglo-Norman lord might come along at any moment and chastise me for being there. But I just met locals out walking their dogs and they seemed friendly enough.

Carlingford is a small town, but it's growing. Halfway between the bustle of both Dublin and Belfast, it's so accessible and yet so rural, scenic and quiet. The little town centre, having been around for so long, has narrow one-way streets that make it inviting and cosy, but it's also busy and seems to be thriving. The narrow streets turn this way and that, with an abundance of tea shops and craft shops, clearly catering for the thriving tourism the town easily attracts.

On my way out of town, I took a further stroll out to the remains of the other ancient buildings – the Friary, Taaffe's Castle, the Mint and the Town Wall, not intact, but definitely recognisable. The town is simply overwhelming with the amount to see and do in such a small geographical space.

Clearly the town is still a working harbour, and with its water sporting activities I imagine the sea area forecast finds an audience. The eastern side of the country generally fares a little better in terms of weather conducive to water sports on the sea, and the small craft enthusiasts in particular spring to mind, the climate generally being temperate and the winds that bit less of a challenge to those starting out. It rarely gets very cold either, so sailing can theoretically go on all year round.

For example, and as I mentioned earlier, it rarely snows in Ireland – it rarely snows big, I mean. Snow flurries, snow showers, a little dusting here and there and some snow on hills and mountains, sure. But full-on, picture-postcard, Christmassy scenes are the exception rather than the rule. However, when it does snow, when it does snow proper, well, Carlingford Lough is one of the places you're likely to find that picture postcard. And, of course, there's a reason for that.

Too cold to snow?

How many times have you heard that? A sage nod of the head, arms crossed over the chest. Everyone knows it: it's too cold to snow.

Because we don't get a huge amount of snow in Ireland, no one here really has any business being an expert on the weather conditions that are, or are not, conducive to snow. Of course, occasionally it does snow. Probably every year we could find reports of snowfall at some or other location on the island, most commonly on high hills and mountains, sometimes at lower levels in the north of the country, but less frequently at the surface further south. But it certainly does happen.

Every now and again there's even significant snow-fall, but these events are so rare they actually become famous in our weather memories, like 1982, 2010 and, of course, Emma in February/March 2018.

And as for cold? Well, it doesn't get that cold either. Thanks to the warm Gulf Stream and the prevailing southwesterly winds that come over it, our climate is very much appropriately referred to as 'temperate'.

And yes, it can get cold, and it has got cold. The lowest official temperature in Ireland, -19.1°C, was recorded at Markree in Co. Sligo on 16 January 1881. More recently it was really, really cold in November and December 2010, and that cold spell was unique because it was prolonged too, and the country ran out of salt for gritting the roads.

But did it snow? Well, yes, as a matter of fact, it did. Because despite the sage wisdom of the public of Ireland, it is actually never too cold to snow.

Snow, of course, is simply frozen precipitation. For it to snow, the precipitation (usually referred to here in Ireland as, forgive the obviousness of this – rain) has to freeze. We interchange the word 'precipitation' with 'rain' here in Ireland because almost all the precipitation that falls over Ireland is in the form of rain. But it includes rain, drizzle, hail, snow and sleet – which is a mix of snow and rain. And there's also graupel, the kind of hail/snow mix mentioned earlier and also known as soft hail or snow grains.

To get really cold conditions in Ireland, we generally need anticyclonic conditions to become established during the winter months of December, January and February. That clears out the air of clouds, and the clear skies at night allow any heat in the earth to radiate away into the atmosphere. The longer the anticyclonic conditions prevail, the colder it gets, certainly in the depths of winter when incoming radiation from the sun is minimal because of the depth of the atmosphere – due to the angle caused by the tilt of the Earth – over the northern hemisphere.

Of course, the other thing that happens when anti-cyclonic conditions prevail is the blocking of weather systems off the Atlantic – the diversion of our prevailing westerly wind that keeps us warm. Not only does the westerly wind bring the temperate weather, it also carries moisture off the sea.

In anticyclonic conditions the air gets drier. If the air is drier, if there are no clouds, then there's simply no precipitation to freeze. The difference is between correla-tion and causation. It doesn't not snow because it's too cold. It typically doesn't snow because there's no moisture to freeze. Introduce a little moisture, and puff – snow. No matter how cold.

So picture the scenario. Anticyclonic conditions have set up over the country, and night after night we lose all our heat as it radiates away into the atmo-sphere, never quite catching up during the day, so each day gets colder than the one before. It's dry and clear and cold. Too cold to snow, even. Then the position of the anticyclone starts to shift, and an easterly airflow becomes established over Ireland.

Here on Carlingford Lough, with the medieval buildings placing the picture-perfect backdrop at the ready, we see those cumulus clouds forming over the sea and they start to become deep- and dark-looking – quite threatening, indeed – as if they were forming some sort of precipitation.

The easterly wind drifts the clouds our way. As they approach, we see them darken the horizon and we get snow. In the US Midwest, off the great lakes, they call this 'lake effect snow'. The cold air travels over the water source – Lake Michigan, say – and dumps snow in heaps over Chicago. Here in Ireland it's the Irish Sea that's the water source. And my, doesn't Carlingford look pretty underneath all that snow.

HOWTH HEAD

53.377298, −6.056470

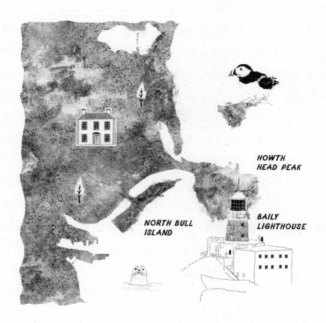

HOWTH
HEAD PEAK

BAILY
LIGHTHOUSE

NORTH BULL
ISLAND

Heading along the coast northwards from Dublin takes you to the Howth Peninsula. The sweeping vista across the city from the Hill of Howth makes it feel like you're in another world entirely from the bustling city. It's best to stay on the path if you hike up the hill, as there's gorse and wild shrubbery underfoot and the cliffs can be dangerous. The Baily Lighthouse stands on an outcrop of rock on the southern side of Howth.

Of all the headlands mentioned in the sea area forecast, Howth Head is probably the most accessible. It neatly divides the coast of the Irish Sea in half, and the DART terminus is just a few steps from the harbour wall. Bars and restaurants await the lucky traveller to provide sustenance for those going to tackle the climb to the summit. And because I live not far from this headland, there was no need for me to head out on a long drive to get there for a rare visit – in fact, the observations that follow have been gathered over a longer period of time and frequent visits, thanks to proximity.

The Howth Peninsula juts out of the coast to the north of the bay, and if you try to access it by road you have to pass through Sutton Cross. Sutton joins the Howth Peninsula to the mainland and there are beaches on either side. A visitor can head south at the cross to take the long, slow route to the summit, passing Dublin Bay on the right. They can take the road on the left to go direct to Howth, with Burrow Beach hidden behind houses that back directly onto that secluded little treasure trove.

Once in Howth proper, you can head along the harbour and start the steady climb towards the Hill of Howth. There are many different paths around, up and over the Hill of Howth, and there are various paths to choose from depending on your fitness level. The views of the city to the south and the beaches of Portmarnock and Portrane to the north are stunning. The paths can get tricky in bad weather, and there have been many fatalities over the years. The fact that it is so accessible makes it no less dangerous.

Although there is a small car park near the start of the climb, it really is better to take public transport to Howth, particularly in fine weather. There's only that one small road to and from the peninsula that bottlenecks at Sutton Cross, so you can find yourself well and truly stuck if unfortunate enough to have taken the car.

Bikes are very popular around the Hill of Howth, as the steep climbs to the summit heading out of the town offer a challenge to the sporty, and the long, steady descent towards the city provides a chance of speeds that would make your hair curl.

The walk around the hill itself can be very peaceful once it is off season and you are off the beaten track. It has several summits, the highest being 171m, and there are routes down to little beaches along the way. In fair weather there are always kids jumping from alarming heights into the seas below. Terrifying.

The rocky outcrop of the Hill of Howth has been the scene of many wrecks. Despite the deceptively gentler waters of the Irish Sea when compared with the Atlantic, ships heading to the busy port of Dublin have been lost there.

When the City of Dublin Steam Packet Company's ship the *Prince* ran into the cliffs just north of Baily in 1846, it was decided that a fog bell would be added to the lighthouse. Before they got around to it, however,

80 more lives were lost when the PS *Queen Victoria* was wrecked in February 1853, and the bell was finally installed in April 1853. There has been a lighthouse on the hill since 1667, and part of that original structure can still be seen where the hugely impressive structure of the Baily Lighthouse now stands.

With views across Dublin Bay, the city and the suburbs, the area would make for a great viewing spot for storm-watchers, if storms were as frequent on the eastern side of the country as they are in the west. But they're not. Storm-force winds and heavy persistent rain are not regular phenomena around these parts. So when a storm does come through, it tends to be remembered. Charley was one such storm.

CHARLEY, AN EXCEPTIONAL STORM

Although in these parts remembered more for the rain – or more precisely the flooding – Charley was born on 11 August 1986, when it was first identified as an area of thunderstorms in the southeastern area of the Gulf of Mexico. The National Hurricane Center watches all areas of thunderstorm activity in the north Atlantic during hurricane season and mid-August is nearing peak hurricane season. By 13 August, Charley was on their radar, no pun intended, and Charley was a developing sub-tropical non-frontal low-pressure system.

The system moved northeastwards through Georgia and South Carolina as a welcome visitor whose rainfall relieved drought conditions at the time. But by late on 15 August, Charley had graduated all the way to Tropical Storm Charley – next stop, hurricane. Sure enough, a ring of thunderstorms circulating the centre developed into an eye wall (the area immediately outside the eye of a hurricane or cyclone), and at midday on 17 August, off the North Carolina coast, Hurricane Charley emerged.

The hurricane-force winds drove a surge, causing tidal flooding, knocked down trees in North Carolina, and left 110,000 people without power in Virginia. Traffic accidents attributed to the storm took the lives of two people. The storm moved northwards along the northeastern seaboard of the United States, continuing to cause damage, and three people died in a plane crash in Maryland, also attributed to the storm. One person drowned in Newfoundland, and the storm was estimated to have cost 15 million US dollars at that time.

After 48 hours as a hurricane, it started to move across the north Atlantic and make the transition to an extratropical depression, the ones we see regularly in our area through the season. Still, it maintained some of its tropical characteristics, like high moisture content, which can transpire to heavy rain and flooding. Now, being sucked up in the westerly flow, Charley started to intensify again on its way to Ireland. In fact, as an extratropical cyclone, Charley had a deep low centre of 980hPA, deeper than its lowest pressure as a tropical cyclone.

Meeting more encouraging atmospheric conditions on the way, the remnants of Hurricane Charley accelerated towards Ireland, with the centre of the storm passing just to the south on 27 August, 16 days after having first been identified.

With its approach, storm warnings of expected heavy rainfall and strong winds were issued. The rain was heavy across the whole of the country, but this time the one place that is usually afforded the quietest life when it comes to rain was possibly the worst affected.

Dublin, the capital of Ireland and the most populous city in the country, is built on the River Liffey. It is surrounded to the south and west by the Wicklow and Dublin mountain ranges. With prevailing southwest winds acting as a meteorological umbrella, Dublin enjoys something of a charmed life in terms of the rainy days that many associate with the climate of Ireland.

The southwest winds forced over the high ranges of the Wicklow Mountains cause the water vapour in the air to condense, and the resulting clouds shed their rain on the windward side of the mountains, leaving Dublin, on the lee side, in the warmer and drier air that remains.

But with storms, the heaviest of the rainfall is always to be found to the north and west of the centre of the area of low pressure. With Charley tracking to the south and towards England, Dublin was directly in the firing line. On the border between Dublin and Wicklow, at Kippure, 280mm of rain was recorded, more than 200mm of that occurring in one 24-hour period, setting the highest daily rainfall total for the country and almost overflowing the reservoir in Bohernabreena.

The resulting flooding in the Dublin area was described at the time as the worst in history. Two small rivers, the Dodder and the Dargle, overflowed their banks to a depth of 1.5m, causing a thousand people to flee their homes. The storm caused at least 13 deaths, 4 of which were drownings in flooded rivers. One death happened when a person was being evacuated from flooding.

The clean-up operation cost more than 6.5 million pounds and went towards repairing roads and bridges destroyed by the floodwaters. But the longer-term legacy means that the words 'the remnants of hurricane ...' sends a current of apprehension through the minds of those who were around when Charley came through.

But while we're here on the east coast and hanging around Dublin, it's a perfect opportunity to talk about the beaches around Howth Head, or, more specifically, Portmarnock. I'm biased – I live in Portmarnock, so I have access to this incredible amenity all year around and do not take it for granted.

The Velvet Strand is a wide sandy beach, 5km long and edged by dunes harbouring protected species of wildlife. It's popular with walkers all year round, and we locals marvel every day at the swimmers braving the sometimes long walk out to the water – it's maybe not quite as wide a tidal difference between high and low tide as on the other side of the bay at Sandymount, but it can be a long walk all the same.

I swim here myself in the summertime. Although the temperature of the sea along the east coast of Ireland rarely drops below 9°C even in the depths of the coldest winter, the air temperature on the way into and out of the water needs to be accounted for, and besides, 9°C is about 28 degrees less than body temperature. No thanks.

The dunes are protected and are at times fenced off from the public. Various initiatives have been tried over the years to introduce natural or environmentally friendly ways to support the dunes, including piling up old Christmas trees – this is done in a coordinated manner, so don't go along and dump your tree on the beach.

The views out to Lambay and Ireland's Eye on a clear day are fantastic. Watching cumulonimbus clouds form over the Irish Sea is a thrill. As well as your run-of-the-mill swimming, there's also a diving club nearby, and people windsurf, paraglide, sail, canoe, stand-up-paddle – any number of beach- and sea-related activities go on here.

And, surprisingly enough, the wide sandy beach has, in the past, made it an ideal spot for aviators. Despite the existence of a monument to the event right on the grass at the entrance to the beach, it's a little-known fact that the first east-to-west transatlantic flight took off right on this beach.

In the early 20th century, aviation was taking off – no pun intended – in a huge way. From the time of the Wright Brothers' famous 1903 flight at Kill Devil Hills, North Carolina, what would become air travel grew at an incredible speed. Just over a decade later, by the time World War I was under way, the competition to develop the aeroplane took on exponential growth as warring countries and continents aimed to capitalise on the advantages of being able to fly.

Within another couple of years, the next race was on, to fly across the Atlantic. With the tail wind as an advantage, the west-to-east journey across the ocean was always going to be the winner, and so it was that the first transatlantic flight landed in Clifden, Co. Galway, on 15 June 1919, piloted by Alcock and Brown and mentioned earlier in this book. The tail wind reduced the reliance on carrying fuel, meaning that the aircraft could be much lighter and swifter.

The first non-stop east-to-west crossing didn't happen until almost a decade later, when a group of German adventurers, joined by the Irishman James Fitzmaurice, took off from Baldonnel aerodrome in Dublin and headed to New York. They landed in Canada, but hey, at least they made it across the Atlantic.

The race continued. There is always something else to achieve, and more development – still at exponential rates – upped the ante further. Charles Kingsford-Smith was a famous Australian navigator who appeared on the Australian 20-dollar note for many years. And with good reason. On 24 June 1930, he flew his plane, the *Southern Cross*, from the Velvet Strand to Newfoundland, and from there onwards to Oakland, California, thus completing the first circumnavigation of the globe traversing both hemispheres. Just a couple of years later, James Mollison, a Scottish adventurer, flew his plane solo east to west, also taking off from Velvet Strand.

Each adventurer was waved off from the dunes by thousands of well-wishers, some splashing holy water on the aircraft. The excitement generated globally by the race in the skies drove development and innovation as well as competition. But the real development came in the decade to follow, when air travel came down from the heights of exclusive millionaire adventurers and into the lives of the general public.

For this, we'd need to move off the beach and into an airport. In the years following James Mollison's historic flight from Portmarnock, a site at Collinstown, in the north of Co. Dublin, was developed, and on 19 January 1940 Dublin Airport opened with a flight to Liverpool. Dublin Airport appears in the sea area forecast as a coastal report station, despite being 8km from the coast. But because it's an international airport, there is a weather observer on-site at all times.

Aviation forecasting

Despite the obvious differences, there are some similarities between aviation forecasting and forecasting for the sea area.

Forecasting for the general public, by the necessity of constraints of time, space and clarity, has some abbreviations in-built (for example, 'cloudy periods' and 'sunny spells'), with no information given or expected on how long a period or a spell is. When forecasting for the sea area, however, and as mentioned earlier, we define how long imminent, soon and later are. With aviation forecasting, we go one step further and give an exact time of when a change is going to occur. Well, exact means within a two-hour window, and in fairness, when we're talking about predicting the future, a two-hour window is pretty good.

Aviation forecasts are also, like the sea area forecasts, for specific areas. With the sea area forecast we divide our area by the headlands that are the backbone of this book. In aviation forecasting the sky is divided into what are called FIRs – Flight Information Regions – and the main forecast is called a Terminal Area Forecast, or a TAF, and is valid for a specific airport.

Here's an example of a TAF:

EIDW 191700Z 1918/2018 24012KT 9999 FEW014 BKN025
TEMPO 1919/2014 25015G26KT BKN012

This decodes as:

EIDW	Dublin identifier
191700Z	The time and date that the TAF was issued
1918/2018	Validity; this TAF is valid for the period from 1800 on the 19th until 1800 on the 20th
24012KT	Current wind direction is 240 and speed is 12 knots
9999	Visibility is greater than 10km
FEW014	This refers to the amount of cloud and indicates height in hundreds of feet
BKN025	Another cloud group – in this case meaning broken (SCT – scattered – is the other group, as well as NSC – no significant cloud)
TEMPO 1919/2014	At times between the 19th at 1900 and the 20th at 1400 the phenomena following the tempo group may occur

TAFs are either long or short, being for either 24 hours or 9 hours.

We have specific language for the sea area forecast, with defined terminology to hand to make decoding the message as clear as possible for the end user. TAFs are a series of codes that take some getting used to but are equally clear to a pilot. The usefulness of a TAF is verified by what is called a METAR – a Meteorological Terminal Air Report. This is an observation that is made at the airport of the current weather conditions. While there is some room for slight change, a valid TAF must align with the latest METAR from the airport.

The sea area forecast has a built-in validator – the coastal reports – but it must be remembered that the sea area forecast is an examination of the weather out to 30 nautical miles off the coast, whereas the coastal reports are, well, on the coast. So there can be some differences. Still, it's necessary to monitor coastal observations – all observations – to make sure that a valid sea area forecast is in line with expectations and to update it if necessary.

An aviation forecaster is responsible for the airports within their jurisdiction as well as the FIR. Warnings are issued for hazards to air traffic while taking off and landing, while in the air, and while parked on the tarmac. Hazards for aircraft include turbulence – turbulence being defined as a rapid change in wind direction or speed – and thunderstorms, either at the airport itself or, if frequent, within the FIR. Cumulonimbus clouds within the region also require a warning, mainly because of the turbulence risk associated with these clouds, and also because of the risk of icing within the cloud. Speaking of icing, that requires a warning too.

Icing at the terminal can cause havoc with airport operations, as aircraft need to be de-iced before take-off. When ice builds up on the aeroplane, it changes the way air moves across the wings, and because aeroplanes lift off and stay in the air due to the difference in pressure caused when air flows rapidly over the aerodynamic shape of the wing, even the smallest differences caused by the disruption of this airflow can be deadly.

The downdraught from a thunderstorm cloud – the cumulonimbus cloud – can be enough to knock a jumbo jet out of the air if experienced during landing, and it has happened. In 1985 a jetliner approaching Dallas Fort Worth Airport noticed an ordinary summer thunderstorm ahead. Not thinking too much about it, they advanced, and as they did so, the thunderstorm intensified. Thunderstorms can be rapidly developing systems, especially on hot summer days.

As the aeroplane flew into the storm, it met first the strong headwind, then the downdraught from the cloud, then a strong tailwind, which robbed the jet of lift and caused it to sink rapidly, slamming down short of the runway. The fatalities on that day in August 1985 amounted to 137 people, but thousands more passengers since then have been saved as the aviation industry implemented new safety measures to enable all pilots to be aware of the dangers of thunderstorms near airports.

FROM WICKLOW HEAD TO CARNSORE POINT TO HOOK HEAD

WICKLOW HEAD

52.960374, -5.998803

Located midway along the beautiful coastline of the garden county of Ireland, Wicklow Head lighthouse is accessed along a road that crisscrosses back and forth over itself to wind its way towards the rocky edge of the Irish Sea.

A round the time I made my journey to this part of the country, abnormally high temperatures over Europe had been recorded through the month of October, with records broken in several mainland countries. Temperatures of over 30°C were still showing up on thermometers in Barcelona when children were preparing to trick-or-treat.

The large blocking high-pressure system set up over the continent kept the usual autumn depressions circulating around the northeast Atlantic. Picking up moisture, the warm ex-tropical air was laden with water vapour, and with each pass, it dumped yet another round of rain or showers over Ireland.

At the start of October, after a relatively dry summer and September, the soil moisture deficits (SMDs) across the country were in excess of 50mm in many parts of the country. The SMD is the measurement (in millimetres) of the amount of water the soil can hold before it becomes saturated. So a deficit of 25mm means rainfall amounting to 25mm will bring the soil in this area to a saturated state. Any additional rain causes the ground to become waterlogged, making flooding increasingly likely.

From a deficit of 50mm at the start of October, within just a few weeks all soils were waterlogged and all rivers were at 'bank full' – that's when river levels have reached, well, the bank.

By early November, after three consecutive weeks with rainfall amounts greater than 200 per cent of the normal for a large part of the country, and up to 300 per cent of the normal in many places, advisories were in place for the risk of flooding associated with continued spells of rain, even though rainfall amounts were not necessarily reaching thresholds that would normally warrant a warning. As a colleague said at the time, a spilt bottle of Evian would cause a flood at this point.

Low pressure was still very much in charge when I decided it was time to take the 80km trip south to visit the next headland on my list, Wicklow Head. Despite its proximity to home, I'd never been to Wicklow Head and only on one occasion to Wicklow Town itself.

Yet another spell of rain had cleared through overnight and there was a signal for a short-lived ridge to follow. That ridge would give sunshine, and the winds would ease too – perfect conditions for a hike. Daytime temperatures were around 14°C, light winds were from a southwesterly direction and, post front, the visibility was due to improve too. This bodes well because, on a clear day, you can see South Stack at Anglesey over in Wales. I hoped I'd get lucky.

The lighthouse at Wicklow Head is about 4km south of Wicklow Town, but there's a parking spot about 2km out the road, and from there a scenic 'path' down past the tiny rocky beaches and up over the hills. I parked up and tightened my boots, preparing for a hike.

The signs in the car park told me it was just a 4km loop out to the head and then back along the road. I ditched the raincoat – the sun was blazing overhead and it definitely wasn't cold.

There were stairs down initially, but the actual path was little more than a muddy rut between the ferns. At one point I came across an American couple helpfully piling a bunch of these ferns over a big muddy puddle in a very Sir Walter Rayleigh-esque manner. I exchanged pleasantries and they assured me that it wasn't much further.

There were warnings along the route advising walkers that there were baby seals on the beach and not to approach them. To be sure, the beach was closed off to visitors, with very clear markings. There were several tourists standing around the cliffs overlooking the seals bathing on the small rocky beach below. The place actually looked a little crowded – the seal beach, I mean. The seals – I couldn't tell from this vantage point if they were babies or fully grown or a mixture of both – were all piled on top of each other. As far as I could tell, though, and in fairness that's not very far, they looked happy and healthy enough.

Looking north towards Dublin the views were spectacular. Wicklow Head juts out into the sea and the views northwards take in the mountains and the bays of Dublin and north Wicklow. The fact that it juts out like this lines it up nicely as a headland to divide out the east coast of Ireland in the sea area forecast. Winds swinging around the south coast can do some funnelling too – like in the north between the Mull of Kintyre and Belfast Lough – leading to stronger winds in the south Irish Sea as far north as Wicklow Head.

I couldn't see Wales, though. The visibility wasn't great and there was a weather front still making its way across the Irish Sea, bringing the cloud base down.

With the amount of rain during the previous few weeks, at some points along the track there was a stream instead of a path, and I was forced up and over into the slippy grass. It was slow going. It wasn't a difficult hike by any means, just slow and at times a bit tedious. I hadn't called ahead, so once I'd reached the lighthouse proper I veered back towards the road and took the shortcut back to the car, where I met the same American couple just pulling themselves back up the steps to the car park. They looked at me as we passed each other, I'm sure thinking to themselves, what the?? They had clearly stopped at the seal beach and then doubled back on the rutted muddy track. I couldn't have overtaken them – it was single file only almost the whole way – so where had I come from? The poor unfortunates had doubled back on themselves and trekked through the ferns.

Two lighthouses, one on Long Hill, the other on the saddle of the head, were set to work on 1 September 1781. They not only marked the headland but prevented mariners from confusing Wicklow Head with Howth Head to the north and Hook Head to the south, both of which had single lights. Over the course of the following century, the higher tower was lowered, as it was always in fog, and the current lighthouse stands down a winding road. It flashes for 10 seconds in every 15, and that's how sailors out on the Irish Sea know they're off the coast of the sunny southeast. Speaking of 'sunny', and while there's no mention whatsoever of temperature in the sea area forecast, this is an appropriate place to take a slight detour and talk about heat.

HEAT, HEATWAVES AND CLIMATE CHANGE

Wicklow Head is bang smack in the middle of what we in Ireland refer to as the 'sunny southeast'. And in fairness, the southeast generally *is* the sunniest part of the country. Its proximity to the continental land mass makes it more likely that anticyclonic conditions, typical over Europe during the summer months, will extend their reach to encompass the counties of Wicklow and Wexford, extending inland to Carlow and Kilkenny at times too.

Worldwide, the hottest seven years on record have occurred since 2015. It's getting hotter and hotter. But the longest hot spell on record here in Ireland happened in the summer of 1976.

The heatwave of 1976 began when high pressure got stuck over Europe from the end of May. It lasted until late August, with the first sign of rain that summer on 26 August. Although much of Europe was affected, the heat and drought were especially severe in the UK and Ireland. Temperatures hit 32.5°C in Boora in Co. Offaly on 29 June, making it the hottest June day on record, but that was the fifth day in a row that temperatures of 32°C were recorded in Boora. By the time it ended, the three-month heatwave had caused the worst drought in 150 years, heightened by the low amounts of rainfall during the previous two winters.

The term 'heatwave' used to be defined as a period of time when temperatures above the average by five degrees were recorded for a period of five consecutive days. But by those rules, we could theoretically have a heatwave in January! Also, given that Ireland has such a temperate climate, when even in our hottest months – July and August – the average daily max is only between 17 and 20°C, we could again theoretically have a heatwave with temperatures of between 22 and 25°C, and calling that a heatwave would seem a little dramatic. So the term heatwave was redefined as a period of time when the highest daily temperatures of 25°C or higher are recorded for a period of five days.

Ireland's temperate climate is maintained by a warm current of the ocean that comes all the way across the north Atlantic from the Gulf of Mexico (that's why that current is so cleverly named the Gulf Stream). Water warms up and cools down at a different rate from solid objects like earth, sand and rocks, and it holds on to heat too – heat doesn't radiate off the water as quickly as it does from solid objects. So, while land masses warm up fast in the summer months, they cool down quickly too. The ocean, not so much.

That warm Atlantic current keeps a nice, steady temperature ranging from about 8 or 9°C when at its coldest, in and around February/March, to the heady heights of about 15°C in the Irish Sea, while on the Atlantic coast of Ireland, the water temperature reaches about 17°C in August/September, and at times a degree or so higher than that too. That's not awfully warm – if you've ever swum in the Mediterranean in the summertime, you'll know that. But it certainly is warm if you take into account the fact that Ireland is situated at 53 degrees north. Compare that with the water temperature off the coast of Newfoundland and Labrador, where it gets down to freezing in the winter and averages out at about 10°C in the summer.

The Irish Sea is at least a couple of degrees colder than the Atlantic, which is a fact that often catches people by surprise. So, with temperatures like that over the Atlantic Ocean, the air coming up over that ocean gets warmed to near those temperatures and keeps Ireland warmer than otherwise. It's warmer in the winter, that is. It's not exactly hot in the summer.

The average daily highest temperature that some parts of Ireland can expect when our normal services apply – that is, southwesterly winds, sunshine and showers, passing areas of low pressure, fronts and such – is in and around 17°C. For temperatures of 25°C, we need a disruption to normal services. We need an anticyclone.

And if you can bear it another minute to get through just a little bit more meteorology, we'll talk briefly about air masses.

Air is made up of lots of gases – nitrogen, oxygen, carbon dioxide, a few other inert gases in tiny amounts, and water vapour. It's easier to visualise how gas moves if we picture it like a liquid. When two liquids have different densities, they don't mix – think oil and water. Think of a salad dressing with balsamic vinegar and olive oil, the one settled on top of the other.

The air is the same. Warmer air holds more moisture, but is also lighter and rises above colder air, which tends to be drier. There are cold deserts as well as hot ones – for example, it doesn't rain in the North Pole or in the Antarctic, because it's too cold. We talked earlier about how it was never really too cold to snow, but when we're talking about the cold deserts, we're talking about a whole different level of too cold – the air here doesn't even have enough moisture in it to form ice crystals in the first place.

'Air masses', then, is a term we give to areas of air where the characteristics are similar. Warm, dry air masses form when the air remains stationary over warm continental land masses – for example, over the Sahara. Warm, moist air masses form when the air remains stationary over oceans near the tropics. Cold air masses form similarly – over Siberia or Scandinavia, for example.

Then, because the Earth rotates, the air masses move. As they move, they bump up against each other. I like to describe it as like a lava lamp, with the areas of warm, moist air being like the bubbles of wax in the lamp, melting and rising as they warm over the hot bulb in the base of the lamp and cooling and falling again when they've moved away from the source of heat.

So it goes for the warm air masses that form over the warm oceans or deserts. They move north and their characteristics change. The warm air rises above that of the cold air masses and we get weather fronts and often-times rain.

There are certain places around the world where these air masses form, and when you think about it, it becomes fairly obvious why. Large land masses like Africa, Siberia and North America give rise to warm, dry air masses due to the large expanses of dry land generating heat during the summer months. The Atlantic and Pacific Oceans give rise to warm, humid air masses.

Areas of high pressure build over these regions and become stable, large and slow-moving or stationary.

Because they're stationary, they take on the name of the region they are attached to – the Azores High or the Scandinavian High are examples. We're familiar with these because sometimes they come our way. Later, we'll talk about what happens when a Scandinavian High moves in over Ireland. (I'll give you a clue for now – it gets cold.) In the meantime, we'll talk about the Azores High.

The movement of air around the Earth is disrupted when it flows over large mountain ranges. There are several here in the northern hemisphere – the Rockies and the Himalayas are just two. These huge, high mountains deflect air up, over and around, making patterns of troughs and peaks in the movement of the air. Sometimes these look like the Greek letter omega (Ω) when isobaric charts are drawn from the resulting pressure patterns.

The Azores High sits in the loop of the omega, as do areas of high pressure in general. Around the edges of the loop, low pressure is deflected and the anticyclone becomes referred to as a 'blocking' area of high pressure.

With the right set-up of weather patterns, the anticyclone usually located over the Azores can move northwards. A ridge of pressure builds over western Europe and the weather starts to become fair.

Initially, high pressure moving over Ireland originating over the Atlantic will bring cloudy but dry conditions for a few days. It tends to be cloudy because of the high levels of moisture in the air that originates over the ocean, but as pressure continues to build, descending air will clear away clouds and the sunshine will improve.

Most heatwaves in Ireland occur in June or July when the days are the longest and the sun during the day has time to build heat. Sometimes, August features higher temperatures, but only for a relatively short period of time.

In 1976, the heatwave started as early as May. Following two relatively dry winters, the water table was already low when pressure began to build and cloud dissipated. With the ground dry it began to heat, and this cycle of heat and pressure-building continued. The anticyclone blocked out the advancing weather fronts off the Atlantic, making rain impossible and leading to drought conditions. Temperatures were reaching higher than 30°C and rising.

The long-term health effects of heat stress caused during the period were noticeable on displacement of the mortality norm – that is, more people died, maybe not initially, but as a direct consequence. Up to 20 per cent more. Heat is, indeed, dangerous.

We've had more heatwaves since then, of course. The summer of 1995 was remarkable, as was 2006 and more recently 2018 and 2022. If anyone is paying attention, they will notice they're coming at us more regularly. As of the time of writing, and according to Met Éireann (the official keepers of the records for the country), the highest temperature yet recorded on the island of

Ireland was 33.3°C, reported at Kilkenny Castle on 26 June 1887.

But there are a few things to discuss here. First of all, Ireland was the last country in Europe to have a national heat record that was set in the 19th century. Indeed, many of the long-standing heat records of the countries in Europe were broken and rebroken and rebroken again in the 21st century. It's beyond doubt that it is heating up out there. But not here?

Well, yes, here too, really. If we were to disregard the heat record of 33.3°C in Kilkenny, the heat record for Ireland (from a dense network of reliable World Meteorological Organization (WMO) standard observation stations) would have been 32.5°C recorded at Boora, Co. Offaly, on 29 June 1976. That is, until it was broken at the Phoenix Park on 18 July 2022, when the thermometer there topped out at 33.0°C, only 0.3 degrees off the Kilkenny record.

There was considerable debate during the summer of 2022 about the veracity of the Kilkenny record, and the debate then turned to the record in Shannon, with some discussion of the positioning of the car park and its black top creating an artificial advantage to the temperature-recording equipment placed nearby.

But while people in areas reputed to hold weather records may not like to see those records challenged, the fact is that extreme and necessary care is taken by weather observers to make sure every piece of data measured is as accurate a reflection of the state of the atmosphere as is possible. Advancing technology has made it possible to hugely increase not just the reliability of data but also the number of data points available. Data are checked and rechecked, and weather observing stations are monitored and maintained to standards set by the WMO.

A recently published paper has had a closer, more scientific look at earlier data. Icarus is the Irish Climate Analysis and Research UnitS. (There may be just one unit, but Icarus flew too close to the sun, and I guess the temptation to round off the conveniently mythological acronym with an 'S' was just too much.) ICARUS is based at Maynooth University and the study in question was published in late January 2023, in conjunction with the Environmental Protection Agency. It used accepted scientific methods, specifically an inter-station reassessment using sparsely available records and making recourse to the new and improved 20CRv3 sparse-input reanalysis product. Yes, that's a mouthful, but that's science for you.

The 20CRv3 is 20th Century Reanalysis version 3. It makes use of a data assimilation system and surface pressure observations to generate a four-dimensional global atmospheric dataset of weather, spanning 1836 to 2015. So that's weather forecasting charts in reverse – weather 'hindcasting', so to speak.

The result? They found no strong substantive evidence to support the veracity of the record at Kilkenny.

In fact, the words 'grossly insufficient evidence' were used. This does not undo the heat record. The report is just one part of one piece of the puzzle, and there are many more. So as I write this, the 1887 record stands, and until Met Éireann approves an official, comprehensive study, there will be no overthrowing of records.

This is as it should be. Every study is welcomed, as each new investigation provides new information. Every question asked in science takes us one step closer to answers, and this is how new scientific methods are found and how new scientists grow.

The whole point of this study, of course, was not just to annoy the people of Kilkenny who own that record, But joking about pride in records apart, the ability to 'hindcast' – to recreate a map of the weather world as it looked prior to industrialisation – is a vital tool in the work of understanding climate change and how changes are likely to impact on populations going forward. As I keep saying, you can't look forward and envisage a future if you don't know where you are now.

The Earth is heating up, and what we used to call 'global warming' has been rebranded as 'climate change'. This isn't a book about climate change, but not mentioning it in a book about the weather would be a classic case of ignoring the elephant in the room.

The fact is that the hottest 19 years on record globally have occurred since the start of this century. Heatwaves and droughts are likely to become more common even where we live. As well as working hard to combat the influence we have on the climate through our behaviour, we will have to adapt to the consequences of heatwaves and droughts and incorporate them into our way of life here.

Like most problems people face, the poor in the developing world will feel the effects the soonest and the hardest, while the rich will be protected for longer by their ability to buy alternatives – expensive food, expensive fuel, homes adapted to rising sea levels or increased frequencies of stormy conditions.

Global climate change will impact those who are directly dependent on the land in regions of the world where extreme weather is already an issue – areas that depend on monsoon rains, for example, will be affected when the rains change in intensity and timing.

The climate emergency can seem daunting to the individual: *What can I do? This problem is too big for my actions to make a difference.* But the reality is quite different. We can all make changes in our homes – small things such as using energy-efficient light bulbs and reducing waste. I commute by bike and I know friends in the countryside who have almost entirely switched to solar energy to power their homes. Hopefully, over time, these personal changes can influence the community around us and expand into the public by degrees, until government ministers have to respond by providing the infrastructure demanded by their constituents. Already we can see more readily available and reliable electric car charging spots popping up around the country, like in the car park at Kilkee when I was visiting Loop Head. There is lots to do, but we will get there one step at a time.

CARNSORE POINT

52.174160, -6.362765

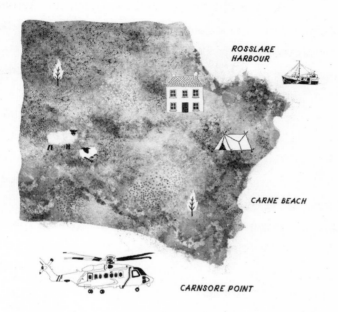

ROSSLARE HARBOUR

CARNE BEACH

CARNSORE POINT

Carnsore Point is the most southeasterly point on the island of Ireland, a good two-hour hike south from Rosslare Harbour. The sunny southeast lives up to its name here, with popular tourist beaches and holiday camping sites dotted around the area.

Carnsore Point is very close to the busy port of Rosslare, and visitors mingle with trucks and those making their way to Wales year-round on the route there. To visit the actual 'point' of Carnsore Point it's probably best to start at Carne Beach a few kilometres to the north, as the roads out towards the point are little more than tracks, not fully paved and broken to pieces with potholes.

The beach to the west of the point is not a swimming beach. It has dangerous undercurrents and sharp drop-offs, the sand is pebbled and rough, and there's no parking. Carne Beach has a small holiday park with facilities, taking advantage of the closest thing on our island to a continental climate. There are also walking routes from the beach out to the point, with a few different levels of difficulty to suit both beginners and those up to more of a challenge.

There's a 4km route that takes visitors out to under the wind turbines that occupy the point. Standing under the turbines might remind you of the visitors from space in *The War of the Worlds* – they're a little bit alien and loom over the landscape. But the sound is soothing. Their gentle swooshing along with the sound of the surf is a relaxing white noise in the background, though granted you might not feel that way if your house were in the vicinity. There aren't many houses in the vicinity of Carnsore Point, though.

Of interest to geologists for the granite – a distinctive pinkish-brown rock – a large area wrapped around the point is considered a marine protected area and an ideal spot for watching seabirds. On the day of my visit, the layers of cloud caught the light, leaving silver slivers of arcs in the sky. In late November there were few birds, and there was just one other person on the beach when I set out – a traveller on his way back to Rosslare Harbour en route home to Wales.

Tuskar Rock Lighthouse is visible from the point, standing 5km offshore to the northeast. The lighthouse gets a mention in the sea area forecast, as the visibility point is read out to complete the coastal report from the buoy at M5, which is actually located about 60km south of Hook Head.

I was very fortunate to bump into Liam Ryan on my travels around this area. He told me he had worked as a lighthouse keeper as a young man. The lighthouses around Ireland have all been fully automated since the 1990s – Tuskar since 1993 – and Liam was stationed for a time on Tuskar. Lighthouse keepers worked a month on and a month off, and while 'on' they were literally on the rock, all the time.

It was a hard and lonely job for a young man. There were no Netflix series to binge on back then, and no PlayStations to occupy idle hands in downtime. The young men worked with their peers in rotas and took their own provisions with them when they were dropped off by helicopter for their month's work, though Liam tells me with a sly smile that the RNLI guys were quite literally lifesavers and would drop them in a box of fags now and again.

Carnsore Point hit the headlines in the 1970s when the Irish government stepped up plans to develop a nuclear energy programme for Ireland and chose the point as a location for one of the first (of five) plants proposed for the country. There was fierce opposition to the proposal, however, and the drive to oppose the development, led by local Wexford environmentalists, gathered worldwide support. The plans were scrapped.

The sunny southeast often sees high temperatures, but around the time of my early November visit, a cold front had moved through. The large contrast in air temperatures at the surface and a sharp change in wind direction as the cold front moved through caused conditions to be just right for the development of a tornado.

That is what residents of a town called Foulkesmill (about a half-hour to the west of Carnsore Point) reported one day, describing how the sky turned black and they heard a loud noise, lasting just a few minutes, at around lunchtime midweek. The damage caused by the tornado has been estimated to be in the region of hundreds of thousands of euros.

A retired doctor and his wife are the owners of a farm boasting a tree-lined avenue and a stunning walled garden. The trees, some hundreds of years old, were ripped from the ground like matchsticks in a matter of seconds. Barns were destroyed and the wall of the walled garden was toppled, the rocks tossed like pebbles. Four cows were killed in the sheds, one instantly and three from injuries sustained during the incident. Buildings were destroyed or seriously damaged, with roofs picked off and blown away. Thankfully the doctor and his wife were not at home at the time, but they were distraught when they returned to find their beautiful home destroyed by this freak occurrence.

But if hurricanes hardly ever happen in these parts of the world, tornadoes are just as rare, so what happens when we have a tornado in this country?

Tornadoes in Ireland

Tornadoes, on the rare occasions they happen in Ireland, are usually on a fairly small scale such as the one I've just referred to. They can occur within deep troughs of weather fronts when the contrast of the air masses is striking and a sharp and sudden change in wind direction occurs. We might call them 'waterspouts' or 'mini-tornadoes', and they're called 'dust devils' in Australia. If you hear a forecast for 'squally showers', be on the lookout and you might just spot one lifting some haystacks or, if strong enough, knocking over a fence. They are more commonly spotted over flat ground and over the sea or lakes, but this could have as much to do with the fact that the view of a forming tornado isn't blocked by trees, buildings or hills.

The tornadoes that we're all familiar with from childhood cinema – the big ones that appear in the movies such as *The Wizard of Oz*, where they take a whole house away, or in the wonderful *Twister*, where they take shiny black SUVs and Helen Hunt's father away – are extreme weather events and are most commonly experienced in Tornado Alley in tornado season.

Tornado Alley is the region in the United States stretching from Texas through Oklahoma, Kansas (where Dorothy, of *The Wizard of Oz*, lived), Nebraska, Iowa and South Dakota. There are tornadoes in other states, of course, and states bordering these regions are also often included in Tornado Alley, but geographically speaking, this is a big enough area. Tornado season runs from around late spring to the middle of summer when cold air spills down off the Rocky Mountains and meets warm air rising out of the Gulf of Mexico.

Geographically, tornadoes are small features. They pass quickly but with a violent energy, usually within the space of a few minutes, unlike hurricanes, which can last a matter of days or even weeks. Wind speeds of up to an estimated 480km/h have been associated with the most destructive tornadoes. That's higher than the highest wind speed in a category 5 hurricane.

As well as the frightening wind speeds, there's also a drastic change in air pressure associated with tornadoes, and it's this that causes objects in the path of the tornado to be lifted into the air, blasting out windows that can't bear the massive pressure change between the inside of a room and the outside.

Here in Ireland, a sizeable tornado occurred on St Patrick's Day in 1995. It left a widespread track of damage and could have cost the lives of people in its path only for the fact that it occurred on our national holiday when children were not at school. Much damage lay in its wake, with several trees knocked down and houses damaged. Livestock were killed when a barn was destroyed.

The following is an account of the St Patrick's Day Tornado of 1995 taken from the Met Éireann records. It describes events in Co. Meath, some distance away from the location described earlier in this chapter, but a good reference nevertheless:

On the 17th of March 1995 Ireland lay in a very unstable, cold strong and gusty west to north-westerly airflow associated with a deep depression situated between Scotland and Ireland. There were rain showers from early morning in many places, with thunderstorms reported at both Mullingar and Shannon Airport before midday. Showers of rain, hail or snow became more widespread and intense as a trough moved eastwards across the country and thunderstorms were fairly widely reported during the afternoon. The tornado which affected county Meath was recorded between 1410 and 1430 UTC, moving quickly eastwards in the strong westerly flow in a practically straight line over a path approximately 18 miles long. It was accompanied by lightning and hail stones of up to 5cm in diameter. Wind speeds near its centre are impossible to determine accurately, but from the level of destruction caused must have been in excess 100m.p.h.; the noise of the approaching tornado was described as 'like the roar of a low-flying aircraft'. The tornado's radial extent was probably between 40 and 60 metres.

There are reports of tornadoes made to the weather office in Met Éireann several times a year. Usually, damage is minor, but occasionally large trees or structures can be damaged by more violent manifestations.

As is the case with thunderstorms, it is impossible to forecast exactly when and where a tornado is likely to occur, and the best we can do is forecast that the atmospheric conditions are in place that make the development of a tornado possible.

But back to the 'sunny southeast' that has been the main subject of this chapter so far, and to exceptional weather once again. Despite the nomenclature associated with it, this end of the country turned out to be one of the most severely affected by a comparatively recent severe cold spell, Storm Emma.

Storm Emma is remembered in Ireland as the snow event of the 2010s. Roads were shut and towns were practically cut off, drifts stood metres high, and school-free days were promised in advance and then delivered. But Storm Emma was only a small part of the story of that cold winter week in 2018. The whole set-up had actually started weeks previously, very high up in the atmosphere.

The layers of atmosphere around the Earth act to protect us from incoming radiation and allow life on Earth as we know it to flourish. The air thins the further from the surface of the Earth we go, and the effects of gravity diminish.

Further up in the atmosphere, the ratios of the components of the air change and there are more or less of certain elements – the ozone layer being a prime example (the clue is in the name).

The layer closest to Earth is called the troposphere and that's where most of the weather happens. The next layer up contains the ozone layer, where the temperature actually rises for a little while due to the activity of the sun's radiation on oxygen molecules – this layer is the stratosphere.

All these layers have differing densities and characteristics. The impact or influence of the Earth's surface is only really notable in the lowest parts of the troposphere.

We call this the 'boundary layer'. The further away from the surface of the Earth, the less influence friction has on wind strength, for example. Friction – from buildings, forests, mountains – causes the wind to divert and slow down. Above the boundary layer is the free atmosphere. It's also colder up there, much colder.

The air circulating around the planet tends to go from west to east as the Earth rotates on its axis. In the weeks preceding that snowy week in February 2018, a thing called 'sudden stratospheric warming' occurred in the upper atmosphere that caused that west-to-east movement to reverse.

Sudden stratospheric warming is a large-scale event that gives off signals in advance that can be readily caught by the computer models. However, not every sudden stratospheric warming results in the reversal of the jet stream. This time, the cold anticyclone over Scandinavia was met with a low-pressure system west of Portugal, establishing the very cold easterly airflow over Ireland.

It had been cold and getting colder for several days, with the high-pressure system landing over Ireland and introducing really cold, clear air. Blocking the advancing weather fronts from the Atlantic that usually keep us warm and cosy as the days went by, it just kept getting colder and colder.

A northeasterly airflow began to send showers over the Irish Sea, and when they fell into the colder air over Ireland, they fell as snow. Snow showers were frequent and widespread over Ireland from Monday, and snowmen were being built with vigour by delighted children everywhere. But still, the forecasters were warning of real trouble ahead.

The country shook its collective head in disbelief that yet more snow was on the way. In fact, in an interview with a very well-respected journalist from a very well-respected newspaper of the day, one forecaster (okay, me) warned that there would be buckets and buckets of snow when I was asked by Ronan, 'How much?' – and then I elaborated to say we were talking metres rather than centimetres. Still, the country was incredulous. This just couldn't happen here.

I was presenting the weather on the Sunday ahead of that snowy week and advised people to look in on their elderly neighbours to make sure they had enough food in for a few days. Warning that the cold spell was going to continue and that snow was going to fall, I was thinking it would be nice to have people who needed help looked after.

The next morning, as I often did, I went to the supermarket to do my shopping. The first thing that struck me was how busy the usually quiet local supermarket was. The car park was full, the shop was packed, and there were no vegetables or fruit on the shelves. Or very little. The shop assistants were busy stacking shelves, but another few metres into the shop and I realised there was no bread either. I started to feel a little bewildered.

I turned to a frazzled assistant and asked what was going on. He informed me that apparently there was going to be a huge storm and everyone was buying in bulk. Stocking up. There wasn't a sliced pan left on the shelf.

Uh, oh is what I thought to myself. 'Really?' is what I said out loud, all innocent and starting to feel a little sheepish. 'Yeah,' he said. 'Apparently, someone said on the news last night …'. 'It was HER!' I heard. 'SHE said it.' I looked towards a middle-aged woman clutching a bag of easy-peel oranges and pointing an accusing finger across the cauliflower at me.

I stood there in the middle of my local supermarket with the eyes of neighbours (and former friends) directed my way, blaming me personally for the absence of Brennan's sliced pans. I scuttled to the till, paid for my purchases and shot home.

We warned that snow was coming the likes of which the country hadn't seen since 1982. And it came. The blocking high over Scandinavia stayed in place and we got colder and colder. The showers streamed in off the Irish Sea – christened 'streamers' by the avid snow lovers in the online community. A new phenomenon gained notoriety that week, and also got a new name: 'the Isle of Man Shadow' made news when parts of Dublin got enough snow to build igloos and some parts missed any entirely.

This happened when the warmer sea surface temperatures of the Irish Sea – the air was frigid by this time – caused convection over the surface of the sea. Convection is the movement of warmer air rising through colder air. When this happens, the water vapour in the colder air cools and forms shower clouds. When the wind was coming from the northeast, it moved over the Irish Sea, and convection caused shower clouds to form, but in parts the path met the Isle of Man.

The Isle of Man, although a small island, was just big enough to halt the convection process, and the air moving over the cold land of the island lost the 'heat' picked up on the Irish Sea. When the same parcels of air moved towards Dublin, there wasn't enough time to re-form the same shower clouds, and the parts of Dublin that the drier, clearer air flowed over got no snow, much to the consternation of the inhabitants in the surrounding areas. Surely this was a witch at work! Nope, just simple science.

The showers continued to gather through the course of Monday, Tuesday and Wednesday, but there was still evidence of a lot more to come. And this wasn't the 'bad' weather yet. In the meantime, an area of low pressure well to the south of Ireland was deepening and bringing heavy rain to the coast of Portugal. The Portuguese Met Office named this storm Emma.

During the course of Wednesday 28 February, Storm Emma took its expected track north. The bitterly cold air over the country met the relatively warm, moisture-laden weather fronts associated with the storm, and the first of the persistent snow arrived on the south and southeast coasts on Thursday evening.

As the low-pressure system pushed north against the large blocking anticyclone, the winds associated with Storm Emma reached gale force. These strong winds penetrated inland across south Leinster, bringing blizzard conditions as the snow blew through the province in drifts. On Friday morning, residents across south Leinster opened their bedroom windows to find snow piled up outside. Their *first-floor* bedroom windows. Cars disappeared under drifts, roads were indistinguishable from hedgerows, and towns and villages were cut off.

A countrywide red warning was in operation for that night as extreme low temperatures and high winds combined to make conditions dangerous for a time across all areas of the country, even where snow wasn't falling.

Making little or no progress northwards against the blocking Scandinavian High on Friday, Storm Emma became slow-moving, and more bands of snow swept across the country. The position of the storm to the southeast of Ireland meant the east and southeast of the country were worst affected in terms of snow amounts. The whole country was still freezing.

Slowly during Saturday and Sunday of that week, as milder air finally started to make its way northwards, the snow turned to sleet, then rain, and a gradual warming and melting occurred over the next days.

During this storm, more than 100,000 homes lost power and 10,000 lost water. When flights were cancelled and public transport came to a standstill in places where roads were impassable, thousands were left stranded. The farming sector was hit badly when polytunnels collapsed under the weight of snow and barn roofs collapsed with the loss of livestock and feed. Food shortages were reported when farmers were unable to get their produce to market.

And yet …

Even with all this disruption and inconvenience, there are still large swathes of the population that still eagerly ask 'Any sign of more snow?' As soon as Halloween is behind us, the most common question I hear is 'Will we have a white Christmas?' From 26 December that changes to 'Any sign of snow? and from Paddy's Day onwards the number one question a working meteorologist will hear is 'Are we in for a nice summer?'

HOOK HEAD

52.123956, −6.929808

DUNCANNON

DUNMORE
EAST

HOOK LIGHTHOUSE

*Still in Co. Wexford, the Hook Peninsula points
like a finger across the bay to Dunmore East in Co.
Waterford. At the very southernmost tip of this neatly
farmed green stretch of land lies the massive and
impressive lighthouse at Hook Head.*

An exceptional point in space and time, Hook Head is on the western edge of Co. Wexford. Across the estuary where the three sister rivers – the Barrow, the Nore and the Suir – meet the sea, is Co. Waterford, and the ancient towns of Waterford, Dunmore East and New Ross are nearby.

The peninsula is sparsely populated, and on the drive towards 'the Hook' Loftus Hall is on your right. An imposing house once described as the most haunted house in Ireland, it was formerly a hotel, a castle and a convent. The house boasts a chequered history and a staircase to match the one on the *Titanic*.

But as you approach the tip of the hook, all eyes are on the lighthouse. The Hook Head Lighthouse is the second oldest continuously operating lighthouse in the world. And it's impressive. The goosebumps will rise on your arms as you take in the medieval building and the history associated with it.

Wherever a large river meets the sea, you can expect turbulent waters and dangerous undercurrents, and the same goes when two distinct bodies of water meet. There are dangerous ocean currents in many places around the world, famously the Cape of Good Hope at the southernmost tip of the continent of Africa, where the Atlantic Ocean meets the Indian Ocean, and which has been the most dreaded part of any sea journey on the way from Europe to the East throughout history. Cape Horn, where the Atlantic Ocean meets the Pacific

Ocean, was equally dreaded as the most turbulent part of the journey by sea from the east coast of America to the west – until they built the canal through Panama, of course.

And then there's the Hook, where the three rivers I mentioned join to enter the sea at a place where St George's Channel meets the Celtic Sea and the Atlantic Ocean. They call the area of the sea that the Hook Head Lighthouse looks upon as the 'graveyard of a thousand ships'; there are many, many more. When Britannia ruled the waves, this area of water was one of the busiest shipping channels in the world, with ships carrying people to and from England and Ireland towards the new world and trading ships bringing riches from the East and convicts to Australia.

When I arrived at the Hook, waves were overtopping and spraying sea foam onto the road. At the top of the lighthouse, it was nearly impossible to stand upright, and painters trying to paint another coat of white onto the beautiful structure were getting more paint on their faces than on the wall in front of them. The attendant down in the cosy tearoom reported that winds were being measured at 36kt, gale force 8.

Another attendant reported that boulders the size of beach balls had been tossed up on land by the sea when Ophelia hit the area and that the waves overtopped the road, extending into the fields beyond the walled area around the lighthouse. And although Ophelia was a terrible storm, it wasn't the worst experienced at Hook Head. (They didn't share which was.)

The tours offered by the attendants at the lighthouse are a must-do. The historical knowledge and delivery by the enthusiastic Jon Pearse is an hour of entertainment and education that should probably be on the Junior Cycle curriculum, if only such a thing were possible. Jon told us that a lighthouse or equivalent has been on this site since the fifth century, when Dubhán, a monk from Wales, landed in Wexford looking for a life of solitude. He established a monastery just a kilometre or so up the road and soon became aware of the dangers posed to passing ships by the treacherous water off the headland.

The headland (Rinn Duáin) takes his name. While it can be imagined, maybe with a little head-tilting and squinting, that the peninsula itself is the shape of a hook, it's actually a coincidence that the name Dubhán translates roughly in Irish to fish hook (duán), and after the Anglo-Norman invasions the area took on the name Point of Hook, or just Hook. But the name certainly suits the headland. There's a certain majesty and strength that just stands there. The structure seems to have a pride that demands a strong name to hold it. The Hook.

Jon Pearse is married to a fisher who is the daughter of a fisher and descended from a long line of fishers who have made their living on these dangerous waters. There is an obvious and valid interest in safety at sea, and the sea area forecast is critical to that. Jon also has a huge amount of pride in his collection of historical facts relating to the region.

He began by telling the story of Dubhán and how the monks started with a fire from locally available materials, then quickly realised that there was a material that made much better fire back home in Wales. So Dubhán went back to Wales and brought back a shipful of coal.

He also realised that his fire was useless in fog, and there's plenty of fog on the south coast of Ireland, especially sea fog during the early summer and late spring months when the sea is still cold but warm air travels up on southerly winds from Biscay. The fog can stay all day; the fog can stay all *week*. When in fog, Dubhán and the monks who came after him would walk along the rocks, ringing bells to warn those at sea that there was a headland and that they were near to the shore.

The monks built the first lighthouse, and when the Vikings invaded, establishing the town of Waterford further up the estuary, they left the monks as they were. The Vikings were seafarers and knew only too well the need for safety at sea and warnings on coasts. If they had bothered to attack the monks who were doing the job of protecting incoming ships, they would have had to replace them with their own men at some point. So the vigil at the Hook went on.

Jon skipped forward in time to tell me about William Marshall, the Lord of Leinster responsible for the building of the tower that stands at Hook Head today. His image graces the walls of the second floor where William made sure the needs of the monks were catered for with living quarters within the tower structure itself, which is still intact.

William Marshall was married to Isabel de Clare, the daughter of Strongbow, given to William as his wife as a reward for all his great deeds, including his work on Magna Carta. The history surrounding this is fascinating and I could have listened to it all day, but my main interest in the Hook was what it meant to those at sea.

HOOK HEAD AND THE SEA AREA FORECAST

From the observation platform at Hook Head, Jon showed visitors the aids to safety at sea that have been provided over the years. The foghorn is no longer operational, but the horn itself still stands in the room. It's a giant whistle-like iron apparatus that Jon did a very impressive imitation of, and he told us of a childhood spent with the horn's blast sounding in the background. The sound travelled for miles across communities, informing residents of when it was time to go to Mass, come home from the pub or milk the cows. When it was decommissioned several years ago, the phone in the observation area rang off the hook, no pun intended, with locals letting the observers at the station know that there was something gone wrong at the lighthouse.

The foghorn was replaced by technological advances to match the needs of ships at sea. Radio signals and Morse code evolved into telephone and computer signals, and satellites now send exact locations and weather information directly to the computers of the boats on the water.

The sea area forecast often tells of gale-force winds at sea while the residents inland don't notice even a sheet swaying on a washing line. Winds over the sea are uninterrupted by friction caused by hills, mountains, buildings and trees. A storm passing by to the northwest may bring storm-force winds through Bloody Foreland and yet the residents of Gweedore might sleep soundly through. And a storm could easily blow through St George's Channel while the residents of the Hook Peninsula might not notice at all, even though gale-force winds may appear on our sea area forecast.

And that's precisely what happened in mid-October 1987. On 15 October of that year, Michael Fish, a renowned and admired senior weather forecaster and meteorologist at the Met Office in the UK, in a broadcast going out on prime-time BBC TV, said that a woman had called the BBC earlier in the day, worried that a hurricane was approaching. He made that up. In order to present a story, he made up the call.

It gave him an opening to introduce what it was that he really wanted to talk about. That's part of the process of communicating the broadcast and it's perfectly legitimate. Our role as forecasters is not just to forecast the weather; it is, very importantly, to effectively communicate that forecast to as many members of the public as possible. The message Michael Fish wanted to convey to the public was that there was NO hurricane approaching the UK. And, as it happens, he was right on that specific matter.

Unfortunately, he is remembered best for being so, so wrong. He's remembered as having given the most famously wrong broadcast that was ever sent over the airwaves. People who knew nothing about meteorology, forecasting, weather or broadcasting suddenly knew Michael Fish. He became the most famous meteorologist in the UK. Michael Fish was, and possibly still is, one of the best meteorologists, forecasters and broadcasters who ever worked in the industry.

The Atlantic hurricane season runs from June to November, and that covers about 97 per cent of the storms that form from tropical depressions. Some form in May, some extend into December, and the peak is early to mid-September. There are also plenty active in the Atlantic during October, and on that particular day, Hurricane Floyd was affecting the Florida Keys.

We've covered hurricanes in other chapters – how they form and how they affect the weather here in

Ireland. So this isn't another chapter about hurricanes. This is about another weather phenomenon called a 'sting jet'.

In that infamous forecast, Michael Fish did, in fact, forecast windy weather. There was a small depression developing on the Bay of Biscay. The computer models (at the time just emerging from their infancy) had, as far back as the previous Sunday, predicted stormy weather for the Thursday or Friday of that week. But later runs of the same model didn't conform, and there was some uncertainty about the development of the system.

The first gale warnings for sea areas in the English Channel were issued at 0630 on 15 October. They were followed, four hours later – about the time Michael Fish was appearing on the BBC – by warnings of severe gales. At midday on 15 October, the depression had a low-pressure centre of 970hPa, and 6 hours later it was moving northeast and had deepened to 964hPa. By midnight, it was 953hPa.

Those numbers don't look like much written down on a page, but a pressure fall of more than 1hPa per hour at these latitudes and at that time of year is an indication of a process, mentioned earlier in relation to Storm Darwin, called 'rapid cyclogenesis'.

Low-pressure systems or depressions commonly form on a wave on the polar front. This is usually a slow process over the course of days, with a circulation developing from a weather front and the resulting winds developing slowly. There is little interaction with the layer of atmosphere above the troposphere, where most of our weather occurs.

Aside from the speed of the development, the mechanics of the atmosphere in this weather system were completely different, and part of that difference was the interaction with the much colder air further up in the atmosphere. The rapid spiral of rising air interacted with the much colder, drier air above, which then fell quickly down towards the surface of the Earth, getting caught up in the circulation and then spreading out as it hit the ground, which was felt as a sudden blast of very strong wind. It is this that forms a sting jet.

On the evening of 15 October 1987, while all these mechanics of the atmosphere were under way, timely gale warnings and then severe gale warnings went out to those at sea. The sea area was well covered in terms of alertness to the worsening situation. However, that didn't stop the loss of many small boats, while a Sealink ferry was driven ashore at Folkestone and a bulk carrier, the MV *Sumnea*, capsized at Dover.

Warnings were issued in the night to the emergency services and to the defence services saying that it was likely the civil authorities would need assistance in the wake of the storm. But when most people in the country were heading to bed that night, there had still been no indication in any weather forecast that a major storm was on the way.

The storm made landfall overnight on the west coast of England at Cornwall and rapidly crossed the country through the night. Moving northeastwards, it felled an estimated 15 million trees as it journeyed through the midlands and out to sea via the Wash (an estuary at the northwest corner of East Anglia). The strongest winds were to the south and east of the centre of the eye of the storm, putting London in the direct path of the devastation.

Historic giant broadleaf trees all over England were uplifted. They were still in full leaf, and consideration was given to whether this was a contributing factor to the loss of such a huge number of trees. Thousands were left without electricity, which took up to two weeks to restore, as cables had been lost during the storm.

Wind speeds of hurricane force were recorded at sea and the highest measured sustained wind associated with the storm was 217km/h, recorded at Brittany. The most damage was done in the southeast of England, where gusts of 130 km/h were recorded for three or four consecutive hours. In all, 19 people died as a result of the storm, and it remains one of the worst storms in living memory.

Ultimately, Michael Fish's forecast stating that there was not going to be a hurricane turned out to be true: Hurricane Floyd did not arrive on British shores that night – there was no hurricane. Instead, there was a devastating storm that, due to its characteristics, was not caught in time by the computer models of the day.

Michael Fish was an excellent communicator and a brilliant scientist, but, in the memory of the public, he will always be the man who was caught out by an unfortunate turn of phrase.

FROM ROCHES
POINT TO
MIZEN HEAD
TO VALENTIA

ROCHES POINT

51.795017, -8.250522

Roches Point in Co. Cork lies on the eastern side of the mouth of the harbour, across from Crosshaven on the western side. As in Cobh, pretty pastel-painted houses face the open ocean. Once past Roches Point, there's nothing ahead but open sea.

I took the turn to Roches Point on a dull and dreary winter's morning. The small town, with brightly painted houses looking out to the water facing towards Cork, was hushed, its inhabitants having more sense than to be outside as the rain came sideways on a blustery breeze – apart from a lone golden retriever out giving himself a walk, that is.

At the end of the narrow winding road that leads down towards the water, there are parking spaces and a couple of picnic tables. I (and the golden retriever) had a wander around and looked out across the water towards Spike Island and the harbour of Cobh behind it.

The Cork to Roscoff Ferry, out of Ringaskiddy, makes its meandering way around the island and passes the harbour before entering the Atlantic Ocean between Roches Point and the lesser-known Weaver's Point on the other side of the mouth of the River Lee. I took this ferry years ago with our family on a very different sunny summer's day, and passing Cobh as passengers on the bow of a giant ocean-going vessel felt like a movie experience. Of course, it is well known that in April 1912 the *Titanic* took on the last of her ill-fated passengers at Cobh, but the ship was too big to fit into the small port, so she had dropped anchor at Roches Point.

In 1813 the harbour master in Cork requested that a lighthouse be erected at Roches Point because of the treacherous conditions for ships entering the harbour.

There was a tower there already that belonged to the Roche family. Edward Roche had got tangled up in the wars of France and Italy, had been imprisoned in Naples, and didn't much feel like just handing over his property. It had been built for banquets and the view of the harbour and he'd just done it up. I'm guessing he wanted to party on his release from prison.

After much toing and froing, the property was acquired by the government and the light was erected and operating by 1817. By 1835 it was deemed that the increasingly busy port needed a bigger tower, so it was replaced by the one currently sitting at the top of the small town, along a winding road of more gaily painted houses.

The whole region is a busy tourist attraction in the summertime. There's a lot to do down around this part of the world. Apart from the fact that every aspect affords a view that wows, there are also plenty of beaches with water activities, there's the zoo at Fota, and there's the beautiful coastal town of Kinsale. And there's the famous holiday park of Trabolgan right next door to Roches Point itself.

The city of Cork, in the county of Cork, is built on the mouth of the River Lee. Visitors to the city might be forgiven for thinking there's a plot designed to drive them mad or dizzy because no matter how many times they cross the river, they appear to be back facing it again. The Lee has, in fact, split in two, and the city was built up on the island in the middle.

The river does not flow directly to the sea at this point; it flows first into Lough Mahon, does another bit of meandering around Fota Island, takes a pass at Cobh, and then goes on out to sea with Crosshaven to the west and Roches Point to the east.

The only issue with Co. Cork is that it's so big. There's a lot of ground to cover and there are a lot of places to see. Driving from Roches Point to Mizen Head takes as long as it takes to drive from Dublin to Cork City – because although the distance isn't quite as long, the roads aren't as good. Once distances and driving conditions are accounted for, however, Co. Cork is for sure the highlight of any visit to Ireland's coastal headlands. Just check the forecast first; it tends to get windy.

A prevailing wind is simply the most common wind direction. In Ireland, the prevailing wind is southwesterly. That means, more often than not, the wind is from a southwesterly direction. If you were to take a bet on what direction the wind will be from on any particular date in the future, you could guess southwesterly and have way more than a one-in-eight chance of being right (the wind is described on an eight-point compass).

This has many and varied effects on several aspects of life around the country. Towns and villages build up in areas because they're sheltered from the wind or take advantage of the warmer conditions in the shadow of a mountain. Dollymount Strand exists because someone built the North Bull Wall to deal with silting problems, and the prevailing wind driving the movement of water diverted around it, carrying sand up the Irish Sea and depositing it on a little bank – and hey presto, a new beach. It's still growing, by the way.

Anticyclones build in the same familiar regions around the world and weather patterns are variable, but the very word 'pattern' suggests they follow a theme. They're predictable, after a fashion. The bigger the weather pattern, the better we're able to predict it.

Part of the predictable pattern is the jet stream and the depressions that form as a result of its behaviour. The depressions move along the surface underneath the steering higher-level winds, often at just about the same latitude as where we are in Dublin – around 53 degrees north. Cape Clear in Co. Cork comes in at 51.42 degrees north, and Malin Head in Donegal is at 55.38 degrees north. Winds go anticlockwise around the depressions, and that's what gives us our southwesterly winds here in Ireland.

The strongest winds associated with depressions and storms are to be found to the south of the area of low pressure, and the heaviest of the rain is to the north. This gives western- and southern-facing coasts exposure to the strongest winds.

With the depressions come the weather fronts. Ahead of an approaching weather front the wind backs, and when the front goes through, the winds veer. Veering winds go clockwise around the eight-point compass, so an easterly wind veers to the southeast, then south, for example. A backing wind goes anticlockwise – a westerly wind backs to the southwest, to the south, to the southeast, for example. If a depression is to the west, and the winds are then southwesterly as a result, a passing front will change that wind to a south or southeasterly – and remember, the winds are strongest just ahead of the front.

Once that strong southeast wind arrives at Roches Point, Cork City is in trouble. The winds sweep the sea through the estuary and through Lough Mahon onto the river walls in the city, with the potential to cause devastating floods. Steep inclines surround the city too, and this exacerbates the problem as rain runs off waterlogged soils and into the city. Tidal surges meeting rain run-off can cause further havoc.

So as well as being exposed to some of the strongest winds experienced by the country, Cork has to contend with the predisposition of the city to flooding. There have been at least 300 serious flooding events recorded in the city since 1841 when someone started counting. In recent decades, both fluvial – that's rainfall-related – and tidal flooding have caused millions of euros of damage. In 2009, between private and public property, that damage came to €90 million and in 2014 a further €40 million of damage was recorded. Similar events occurred in the winter of 2015/2016 and again in the autumn of 2020. But wind and flooding are only two of the weather features that can affect Cork quite badly.

FOG AND VISIBILITY

The third thing Cork has to contend with to a degree more than other parts of the country is fog. Most of the weather events in this book are wind events, and the most commonly occurring serious weather experienced by those at sea involves strong winds. But there are other hazards. Snow, sleet, rain and drizzle don't feature too much in terms of threat level to most seafarers in this day and age. Obviously, very heavy rain and very heavy snow aren't exactly welcome, but they're not overly common around our coasts although, of course, both happen once in a while and when they do occur, they usually come with very strong winds too, so, fun.

Wind and weather both get their own heading on the sea area forecast and so does visibility. 'Visibility' – if you can't see, you can't see where you're going – is measured in nautical miles for the sea area forecast and described in the body of the forecast using the descriptors 'good', 'moderate', or 'poor'.

Descriptor	Visibility
Good	More than 5nm (nautical miles) (9km)
Moderate	2–5 nm (4–9km)
Poor	0.5–2nm (1–4km)
Fog	Less than 0.5nm (1000m)

At sea, visibility is reduced by the presence of water vapour condensing in the air. Visibility can also be reduced by haze or smoke, but by and large, it's going to be water vapour. When a certain amount of water vapour has condensed to reduce the visibility, but not so much that it's less than 1000m, that's called 'mist'. When visibility drops below 1000m because of water vapour condensed in the air, that's called 'fog'. Clouds are made of condensed water vapour, so basically fog is a cloud with no legs.

Warm air holds more water vapour than cold air, so as air cools, the excess vapour suspended in it condenses and becomes visible to the naked eye. Air cools as you move away from the heat radiated away from the Earth, so that's why clouds are up high.

During the day, as the sun shines on it, the Earth absorbs the radiation from the sun and heats up. The ground then radiates off that heat and warms the air directly in contact with the ground. The warmed air rises and is replaced with sinking cold air, and that process of warming serves to warm the layer of atmosphere in which we live.

At night-time, once the sun goes down, that process stops. The Earth is now radiating off all its heat and that heat is not being replaced. There's a store of heat there, though, as it has been getting radiation all day long, so it takes a while to cool off. Depending on the time of year and the amount of sunshine during the day, it can take several hours of radiating away heat before the air at the surface, touching the ground, starts to become cool enough that the water vapour suspended in it becomes visible. In the summertime, it might not happen at all, because before the air is cool enough, up comes the sun again.

The water vapour that does condense out, though, falls onto the surface, just at the surface. We see it in the morning and we call it 'dew'. Meteorologists have a phrase for the exact time this will happen. It's when the air temperature meets the 'dew point'. The dew point is the temperature at which dew forms – the temperature at which the water vapour suspended in the air will condense. When the air at the surface becomes saturated but the water vapour stays suspended, when it doesn't fall to the ground as dew, mist forms, thickening into fog.

Around lakes, rivers and even soggy fields, there's more water vapour suspended in the air, so mist and fog form faster and in more abundance. So it makes sense that fog forms quickly over the seas and oceans.

The sea surface temperature around Ireland varies from lows of around 8 or 9°C in the depths of winter – the sea is actually at its coldest in March – and peaks in August at about 16 or 17°C in the southwest. During the early part of late spring and early summer, as warm air from the south moves northwards over our cooler waters, fog forms in abundance and sea fog can become a real issue. The sea fog can drift onto coasts and then inland, and it's very common on the south coast. The rest of the country might be enjoying a warm, sunny day and complaints will be streaming in from Cork City on the south coast – they're mired in fog. A dull, cold day.

The fog will only drift in a few kilometres from the sea and then it'll be burned or dried away by the dry ground underneath and the heat absorbed by the radiating sun there, but for those few people living on the coast, and especially the cities that are right on the coast, that's very little consolation.

There are five things that clear fog: the wind, stirring up the air and blowing away the water vapour; the sun, basically burning away the fog; clouds rolling in over the fog and radiating heat; and air moving in that's either drier or warmer than the air currently suspending the water vapour.

Happily, sea fog is usually – though, of course, not entirely – a spring and early summer phenomenon, and although a nuisance, it's not tremendously dangerous once proper caution is taken. And of course, as is known the world over, when the sun comes out in Cork, there's no place on Earth comparable.

MIZEN HEAD

51.451416, -9.818612

MIZEN HEAD

CAPE CLEAR

Mizen Head is at the furthest point south on the Mizen Peninsula, and in the sea area forecast it divides the stronger winds off the Atlantic west coast from the lighter breezes experienced further east. Rocks and ocean are the main features here, and the Mizen Peninsula must be one of the most spectacular places on Earth.

The journey around the headlands of Ireland's sea area forecast brings us to 15 headlands and outcroppings, and almost all of them are spectacular. But the Mizen Peninsula takes the standard and ups the ante.

It's very far away, though. It's not that it's very far away from Dublin – not at all. The Mizen Peninsula just seems to be very far away from Cork! For example, Google tells us it's about a two and a half- to three-hour journey from Dublin to Cork, but it's then a two-hour journey from Cork to Mizen Head. Nevertheless, I was really looking forward to going there.

During Storm Ophelia, a young girl from Schull wrote me a note on a piece of pink notepaper, decorated with little pink flowers and hearts. The note said, 'Dear Ms Donnelly. Thank you so much for warning us all about Hurricane Ophelia and calling a red warning. My dad is a fisherman and he is safe. My grandad is a fisherman too and he is safe. From Olivia O'Driscoll, Schull, West Cork. PS: Thank you for two days off school.'

Olivia's mother, thinking the note was cute, sent it to a friend who posted it online. It caught the attention – and the hearts – of the public when it went viral across social media and the print media and was then read out on the radio by Ray D'Arcy. The simple sentiment of the note struck chords everywhere. Olivia's dad and grandad relied on the weather forecast, not to know whether it was a good day for a stroll around the zoo or to put the washing on the line. These men relied on the forecast for their lives.

Thanks to the modern ways of social media, I was able to arrange to meet up with Olivia and her family on my way to visit the Mizen Peninsula, and I was treated to a guided tour of the area by the very people we meteorologists are here to help protect. This was always going to be one of my most anticipated visits to a headland.

I got in touch with Olivia's mother, Caroline, now a councillor in the area, and I met with Caroline and her family on a cold winter's morning at the local café. Olivia was being treated to yet another half-day off school at the behest of the weather forecaster. We were joined by her father, Seán, who informed us that he had just come in. At first, I assumed he meant in from fishing and was incredulous that he would be on the water in such rough weather – there was a gale at sea. But no, he'd been surfing! There were shocked expressions all round, but Seán surfs every day and knows the beaches and coves round the area where he can do so safely.

We all chatted for a while over copious pots of tea before discussing the sea area forecast. Seán is very familiar with the forecast and confirmed that he listens to it both as read by Met Éireann forecasters live on RTÉ Radio 1 and as broadcast by the coastguard through Valentia Radio.

Seán shares his knowledge of the rules of meteorology and recites Buys Ballot's law, mentioned earlier: if a person stands facing the wind, the atmospheric pressure is low to the right and high to the left. He's correct, but it takes a few seconds for confirmation at the table. We know Buys Ballot's law as standing with your back to the wind and low pressure being on your left, and switching from left to right and east to west at the same time makes your brain fizz. And, as Seán very quickly followed up, this only applies in the northern hemisphere; the opposite is correct in the southern hemisphere.

This is critical knowledge for understanding the sea area forecast. The first section of the forecast refers to the meteorological situation in which the location of the areas of low and high pressure are given relative to Ireland – for example, a depression of 980hPa is centred 200 nautical miles west of Belmullet. The sailor off Mizen Head then immediately understands that the wind is coming from the southwest and can listen to the rest of the forecast for confirmation of wind strengths and when any changes are likely to occur. It's satisfying to sit and listen to an actual fisher, active on our coasts today, and know that the theory is being put into practice.

Before I headed off with his family for a guided tour of Mizen Head, they pointed out Cape Clear and Sherkin islands a short distance off the coast. These islands are just 10km off the coast at Schull, and there's a weather station on Sherkin. When a storm comes through on the prevailing southwesterly airflow, it's the station at Sherkin that is often the one to register the earliest highest wind speeds. Back in the comfort of the weather office in Dublin, it's often easy to dismiss the high wind speeds of the Sherkin weather station as, 'Well, that's a coastal station and always has much higher wind speeds than those experienced inland.' But there I was, standing at the beach in Schull, and you would nearly see the weather station from there on a clear day. I was very conscious of the busy town and the impact that these winds must have on the people living there – and still more the impact on those little boats in the harbour.

The weather that day was harsh – it was windy and there were heavy showers. The sea surface temperature was relatively high and cold air had moved in to clash with the unseasonably mild weather that dominated the weather charts into the autumn. Thunderstorms rolled in off the wild Atlantic, and ground strikes showed up on the horizon, the rumble following shortly afterwards indicating just how close the storms were. The roads leading out of Schull and around the headland were winding and picturesque.

The scenery around the area was breathtaking, and there was evidence everywhere of holiday villages and homes, locked up for winter. The landscape was hilly and green, and the feeling of remoteness would be inviting to anyone seeking peace and quiet.

The pretty beach of Barleycove is near the last turn to the tip of the headland. This is where the surfers find

the waves. The swell coming in around the headland of
Mizen breaks to the east and the rollers can be ridden
to the beach. On my visit, there was protection from the
gale-force prevailing southwest wind, while the swell that
it generated fed the rollers that international champion
surfers come to ride.

There's a visitor centre open from November to
March, but it's only accessible at the weekend. I parked
up in the car park just as a heavy shower erupted off the
ocean. The downdraught nearly knocked me off my feet
as I ran to the shelter of a porch off the main building,
and we shouted at each other over the roar of the wind
and rain. I was suitably dressed in boots and a raincoat,
but even standing upright was a struggle in the gale-
force winds, and my phone was whipped from my hands
as I tried to snap the obligatory selfie at the end of the
pedestrian access point.

There are fences closing off the area and it's clear
they are needed. The cliffs are high and the waves lash
against the rocks below. There is a pedestrian bridge that
was closed on the day of my visit. The original bridge,
a concrete structure, joined the island to the mainland
from 1907 but was deemed unviable in 2005. A new
bridge was commissioned and was completed in 2011.
Not a huge fan of bridges or heights, I was secretly glad
I didn't have to cross it. Out on the outcropping beyond
our view, the footbridge takes you to the Mizen Head
signal station, described at the visitor centre as the first

radio beacon on Irish coasts, installed on New Year's Day 1931.

My visit to Mizen on a wet, stormy day left an impression of the exposed wilderness of the peninsula. There are no tall trees on the headland. They probably wouldn't last – how could they, facing gale-force winds for almost a quarter of the year? But the beauty of the headland – the rocks, islands, beaches and fields – was awe-inspiring.

I'll go back in the summer when it might be just a tad less windy, though I probably still won't cross the bridge. But for now, it's time to turn our attention to another famous Cork location, a little further off land, where weather also plays a huge part, and where there were catastrophic consequences when a storm hit during a yacht race in 1979.

FASTNET ROCK AND THE STORM OF AUGUST 1979

The Fastnet Race is one of amateur yachting's greatest challenges. Always daring. Always demanding. Always dangerous. Thrill-seeking sailors racing against the elements are challenged by the wind and swell on these waters. Yet nothing prepared crews and rescuers for the historic storm of 13 August 1979.

The race has been held every two years since 1925, with a break during World War II, when the oceans around these parts were a hunting ground for submarines. It's a 605-mile course that starts in Cowes, a seaport town on the Isle of Wight. From there the competing yachts sail westwards through the English Channel towards the Celtic Sea, then head for Fastnet Rock – the turning point in the race – before sailing, exhausted, back to the southwest coast of England to the finish in Plymouth.

Fastnet Rock is the most southern point of Ireland, located 6.5km southwest of Cape Clear Island and 13km from Co. Cork. It is also known as 'Ireland's Teardrop' because it was the last part of Ireland that 19th-century Irish emigrants saw as they sailed to North America.

The rock looks quite daunting, reaching up in bleak blackness from the ocean and the huge waves surrounding it. A lighthouse was originally built there in 1854 as a replacement for the lighthouse on Cape Clear. Because the location on Cape Clear was often shrouded in mist, a location further out to sea was sought, and the southernmost point of Ireland – Fastnet Rock – was chosen. But stormy weather and high waves meant that the exposed lighthouse was challenged over the years. During a storm in October 1881, the whole tower above an earlier reinforcement snapped off and was carried away, fortunately with no loss of life.

With this in mind, and a resolution by the Commissioner of Irish Lights that the lighthouse at Fastnet was not powerful enough for its very important position, sanction was obtained in 1891 to replace the cast-iron tower with a granite one that would contain the most powerful light possible. Now, Fastnet Lighthouse is the tallest and widest rock lighthouse tower in Ireland and Great Britain. It was staffed until 1989, when it was changed to automatic operation.

But returning to the 1979 race ... The tragedy caused by the storm all started as a large depression known as 'low Y' that formed over the Atlantic Ocean during the weekend of 11–12 August. The oceans in the northern hemisphere reach their peak temperature in August: more heat, more energy. Any developing storm showing signs of rapid cyclogenesis is always likely to prove a serious situation.

On 13 August, low Y began to intensify rapidly and turn northeastwards, towards Ireland. The shipping forecast issued by the British Met Office early on the day of the race forecast wind speed up to gale force 8. Gale force is estimated to be the upper limit of what yachts can handle and is considered to be manageable as long as the boats are on the open sea, away from rocks and stronger tidal currents.

By 14 August, the low pressure was centred over Wexford. Land-based weather stations reported gale-force winds. The strongest winds were still out to sea, exactly over the race area. The Meteorological Office assessed the maximum winds now as force 10 on the Beaufort scale; many race competitors believed the winds to have reached force 11. By this time, it was too late for the yachts to return.

Analysis by meteorologists as part of an inquiry into the Fastnet storm showed that storm-force winds with very high seas reached the Fastnet Rock area a little before midnight and moved rapidly east across the race area during the next three hours. Stormy conditions persisted until about midday, and then winds eased somewhat from the west as the storm moved away. During the stormy period, mean winds reached 50 to 55kt at times (the upper reaches of storm force 10), with gusts up to 68kt and waves as high as 50 feet. The peak of the storm happened during the night hours of the race, which made navigation even harder. And with the storm moving so rapidly through the area, conditions were changing all the time. Even for the hardiest, the most experienced of the yachtsmen and women, the scale of it must have been overwhelming.

The air was moving anticlockwise around the centre of the storm, so as the storm moved through the sea area of Fastnet, the wind was constantly changing direction – a term described as a cyclonic variable.

When the associated weather front passed the Fastnet area, there was a marked and rapid wind veer. Wind backs ahead of a weather front, then veers behind it. This particular front was violent and severe, and the trough over the area deep and marked, so this resulted in the wind and waves suddenly coming from a different direction – facing the yachts from a different angle. Those in the vicinity of the Fastnet Rock experienced the veer during the hours of darkness, and, for them, the lack of conformity between wind and sea directions made conditions even more challenging.

Later research into the meteorological details of the storm suggested that a strong jet of cold air came down from above, turbocharging the already gale-force winds and adding up to storm force or even severe storm force 11 on the Beaufort scale in a narrow band exactly over the Fastnet Rock sea area (from D.E. Pedgley [2012], *The Fastnet Storm of 1979: A mesoscale jet*).

It is rare but not unknown to have storms in August. The fact that this storm developed as it did, with a core of extreme winds in precisely the race area around Fastnet, makes it truly a unique and very unfortunate event.

The devastation of the 1979 storm meant that many boats never reached the finish of the race that year. Of the 303 starters, only 86 yachts finished. Of those, 194 boats made what are officially termed retirements, and 24 yachts abandoned the race. Tragically, 18 people lost their lives – 15 participants of the race and three of those sent to rescue them.

VALENTIA

51.909511, -10.354222

Valentia was one of the original coastal stations and has been a constant presence in forecasting. One of the oldest observing stations globally, it is a vital source of scientific information. In 1892 the observatory moved off the island to newly built quarters in Cahirsiveen, but it maintained the name Valentia. In recent years an automatic station on the island of Valentia has taken back the role, and Valentia reports now come once again from the island of Valentia.

There's so much to cover on Valentia Island that it might merit a whole book of its own. I arrived late on the island, and in winter the car ferry that operates continuously from April to October between Reenard Point and Knightstown on the island doesn't run, so I had to drive the long way around via Portmagee and drive the length of the island to Chapeltown.

Never having been to the island before, I had no idea what to expect. I'd been this far south in the past when I'd visited the Skellig Islands – the mystical islands that rise out of the water off south Kerry like majestic, alien ships. They host rare birds such as the puffin and gannet, and I took the Skellig Michael boat tour one breezy sunny summer day with the kids. Out from Portmagee, the speedboat went much faster than we were comfortable with, and we sat in stunned silence, gripping the handrails and staring in awe at the black cliffs rising out of the water. I also visited the observatory at Cahirsiveen. But I'd never actually been on the island itself.

Valentia Island was included in the original coastal reports established by Vice-Admiral FitzRoy in the mid-19th century, and coastal reports were observed and returned from the island from 1860 until 1892. The decision to move the monitoring station back to the island was made when Met Éireann judged that the site at Valentia Observatory in Cahirsiveen had become compromised by increased traffic and buildings in the region.

The island was originally chosen because of its location. It is the first land observation between New-foundland and Europe, and its value, along with that of Belmullet, cannot be overstated either in history or in the present day. For the same reason of location, Valentia was also chosen as the site that would host the start of the transatlantic cable, and the fact that it did host the cable was another reason to put a coastal report there.

The impetus to start a transatlantic cable came about when it was clear that navigation at sea was still not as safe as it could be. The longitude problem had been solved the previous century by Harrison (mentioned already in my account of the storm of 1927 on page 58), but inaccuracies still existed between the longitude from the US versus ships sailing from the UK in Europe, and those inaccuracies were causing ships to be lost at sea.

The transatlantic cable project brought jobs and prosperity to the island, and the town of Knightstown grew up around the workers' needs at the time. Handsome Victorian homes built for the workers on the project still face out towards the mainland in a coveted row. The workers were well looked after, the evidence still there in the tennis court laid out neatly in front of the row of houses. The work on the cable was tough and the island was very remote, so incentives to work there needed to be high.

The project began in the 1850s, when work on the first cable started, and the first successful communication

– a message of congratulations from Queen Victoria to US President Buchanan – was sent on 16 August 1858. Work on subsequent cables – the first one only worked for a couple of weeks – continued until the 1870s.

Because of my late arrival, I stayed the night on Valentia Island. The hotel, the Royal Valentia Hotel, was built in 1833, and the sense of stepping back in time on arrival overwhelms you as you enter a foyer that looks very much like it must have during the era when work was being carried out on the cable, and later on the original observatory.

My first port of call was a hike up to the site where the cable was first laid, out the road that runs the length of the island to the public car park for Bray Head. There's a track that leads from the car park out along the headland towards, well, the end of the world. Or what might have been considered, if not the end of the world, then at least the next island to it.

The path heads over a stile and climbs steadily, and the tower that housed the last outpost of workers on the cable project stands at the top of the head. There's a broken wall, complete with broken gateposts half-surrounding the dilapidated old building, and with nothing but the cliffs and the Atlantic Ocean surrounding the spot for 330 degrees around the site, you'd have to wonder what the builders thought they were keeping out with this low wall. You could see anyone coming up the hill from at least a half-hour's walk away.

I stood on the site and took in the history embedded in the innocuous-looking field, looking out towards the Skelligs. It's impossible to contemplate how the monks who occupied the islands lived, voluntarily stranding themselves in the bleak monastery there. There were monks there for 600 years, eking out a life in treacherous conditions, the first observers of the gales that form the nuts and bolts of the sea area forecast. Turning back towards the island, you get the view of the Ring of Kerry and the mainland. Again, it's easy to bandy the word 'spectacular' around, and it's very easy to bandy it around in West Cork and Kerry.

I headed off on a guided tour. When I went to leave the hotel that afternoon, I had been told there was a gentleman waiting in the bar. I was puzzled, as I didn't know anyone there. But word travels quickly around this small, friendly and welcoming community, and the local priest, Fr Patsy Lynch, had heard that there was a meteorologist from RTÉ and Met Éireann on the island. He was determined that I did not leave without having first appreciated the full array of what Valentia had to offer. Fine by me, I thought. Lead the way!

We headed towards St Brendan's Well and the sign on the road that says: 'CULOO ROCK This area is dangerous at times of heavy swell and strong winds. Anglers are warned not to use the rock at these times.' This is a reminder – on an island jutting into the Atlantic – that the Atlantic rules. Patsy gave a non-stop history of the glory of the island and pointed out the shabby wreck of a cottage that was used in a Guinness ad once upon a glory day. Did we want to stop for a photo? No thanks, and we moved on.

We climbed steeply to the top of Geokaun Mountain. This is private land and there's a fee to access the top, but the road is maintained, there are viewing points and picnic tables, and the fee is small. There are 360-degree views, and even on that winter's day, we could see far and wide. To the north is the Dingle Peninsula. Seaward we can see Beginish Island, a small island that is uninhabited now, apart from two tiny dwellings that may be used as holiday retreats, but was once a 10th-century Viking settlement. Looking down we could see the lighthouse that has been guiding vessels towards Valentia Harbour since 1841, and in the far distance is the Slate Quarry, our next stop.

The Slate Quarry was once the source of the most sought-after slate in Europe. That slate graces the roofs of the Houses of Parliament in London and the Paris Opera House. It was used to make billiard tables, including one for Queen Victoria and one for the Duke of Wellington. Workers here were paid four times the rate of pay of quarry workers on the mainland, the slate was that sought after. Standing at the mouth of the quarry was equal parts terrifying and awe-inspiring. A grotto to Our Lady is perched high above the quarry opening, as if to ward off potential danger. The quarry is active, and thankfully that means I couldn't be dragged inside to have a look around. I don't like caves any more than I like bridges or heights.

We drove down close to have a look at the light-house before heading to the final stop on our whistle-stop tour of the island, one of the most fascinating aspects of the visit – the sub-tropical gardens of Glanleam House. Long before we got there, we saw evidence of the possi-bility of sub-tropical conditions on the island. It was almost December, but I kept insisting that Patsy pull over the car so that I could take photographic evidence of the hydrangeas in bloom all over the island – giant blue blooms on plants so large they act as hedges encircling properties.

As we drove down towards the house and gardens at the end of a winding road, I was stunned into silence at what I saw. Ferns as tall as houses dominated the road. There was a rhododendron still in bloom, and the gardens were full to overflowing with exotic, sub-tropical flora. We were very lucky to be in the driveway when the owners of the property arrived back after picking up their kids from the local school. We'd been in the school earlier and the children had just been telling their parents all about our visit, so it was all a very merry reunion.

We got a brief history of the house and the gardens that were planted 150 years ago, and some of the ferns are that old. They grow an inch every 10 years and are so prolific in these gardens they are self-seeding. I lamented that I haven't been able to keep one alive longer than one season in my own garden.

But my garden is cold in the winter. This garden is *never* cold. I pointed to the mountain that looms over the garden and thought, well, that's why. The much bandied-about term 'microclimate' is used in many a town in the country – each claiming the weather in their town is different, usually better, because of their own unique microclimate.

But towns – and gardens – that sit nestled into the side of a mountain *do* tend to have microclimates, and this garden is sitting under a mountain right on the edge of the warm Atlantic Ocean. The temperature of the ocean doesn't fall below 10°C here, even at its coldest, which is in February. So that means the air travelling over the sea tends not to fall below 10°C. If forced to rise, this air will dry out 'adiabatically'. Another of those very scientific words used to intimidate, 'adiabatically' just means without the addition of an external heat source. The air sheds its moisture on the windward side of the mountain and 'falls' down the garden side several degrees warmer. And voila, ferns the size of houses.

Our trip around the island brought us in a circle back to Knightstown, and it was hard to believe that only a few hours had passed and that the island is that tiny and that huge. A small geographic space holds so much history and is the site of so much scientific discovery and observation.

It was just to the west of where I was standing that early on the morning of 16 October 2017 Hurricane Ophelia, mentioned earlier in this book, made landfall. More accurately, we should refer to ex-Hurricane Ophelia, because Ophelia had transitioned to extratropical storm status just a few hours earlier. So before we come to the end of this book, it's appropriate to describe in more detail just what happened when this remarkable weather event hit Ireland.

OPHELIA

Other storms and ex-hurricanes have recorded higher wind strengths or rainfall amounts, but it is possible that none has ever had such an impact on the society of Ireland as did Ophelia. A red-level warning was issued for the entire country, and schools and many businesses closed. The people of Ireland stayed at home and watched as a hurricane – not the tail end of a hurricane, or the remnants of a hurricane, but an actual hurricane – made a direct course for the country.

The National Hurricane Center (in Florida) had identified Ophelia as a tropical storm on the previous Monday. Earlier still that month, Ophelia first started developing on a trough, as an extratropical low-pressure system. The ocean surface temperature was actually half a degree cooler than is usually necessary to support the development of a hurricane, but this depression had a cold core, and the contrast with the warmer air at the surface was enough to generate sufficient energy for a hurricane to develop.

Meanwhile, to the north, another hurricane – Hurricane Nate – was living out the remains of its days, sending its ex-tropical remnants over Ireland and western Europe and setting up a weather chain of events that brought Ophelia in our direction. Ophelia holds the record as the furthest east a hurricane has ever formed.

Taking a step away for a moment, let me try to describe why, as Ophelia transitioned to an ex-tropical storm that Monday, the wind strengths associated with the storm actually increased rather than decreased as it moved northwards towards Ireland.

The jet stream is an area of strong winds that are observed on average about 30,000 feet above the surface of the Earth. Named for the aeroplanes (the jets) that take advantage of their strength, these winds are the reason why a transatlantic crossing takes considerably less time going west to east than vice versa.

The strong winds occur when cold polar air meets warm tropical air moving out of the tropics. Warm, moist air has a different density from cold, dry air. The two don't immediately mix, and the differing pressures associated with the two cause this pressure to build up – observed as wind. Wind speeds at the core of the jet stream of more than 239kt – that's greater than 440 km/h – have been observed.

The shape of the jet stream is described as being either 'zonal' (when the jet takes a straight-line course from west to east) or 'meridional' (when it meanders south and north in a curving pattern). In the second week of October 2017, there was a very strong southerly jet located just to the west of Ireland.

There can sometimes be a disconnect between the upper atmosphere where the jet stream is located and the surface, but at the time that Ophelia was making its way north towards Ireland, the remnants of Hurricane Nate, drawing up warm tropical air at the surface of its frontal system, was butting up against a cooler anticyclone situated over Europe.

Ophelia was perfectly located to get tangled up in this frontal system. The energy available was enough to intensify Ophelia after it had transitioned from hurricane to ex-tropical storm, and the strong southerly jet stream meant the hurricane travelled 1800km in 24 hours, arriving with all this energy and velocity early on Monday morning, 16 October.

The National Hurricane Center had been tracking the storm for the week, as indeed they track all Atlantic tropical depressions and hurricanes, and had been issuing advisories on the track and expected wind fields associated with the storm. As is the case with any tropical storm, they engaged with the met offices of the countries they expected would be impacted by the hurricane. This was the first time that conference calls were ongoing between the National Hurricane Center, the British Met Office, and the Met Éireann forecast office in Ireland.

The National Hurricane Center updates its website regularly with the expected tracks of hurricanes and tropical storms, and these images were the biggest trending images on social media that weekend. The people of Ireland very suddenly and very enthusiastically became avid storm chasers. The last conference call with the National Hurricane Center took place on the Sunday, as by 0300 on the Monday the transition to ex-tropical storm had taken place. By the time the conference call between the UK and the Irish met offices took place, we were on our own.

All through this period, the computer models had been behaving marvellously. Although there is always uncertainty when it comes to the exact track of low-pressure systems, and small shifts in the position of the centre of a low can make a huge difference on impacts, by the time the high-resolution models used by Met Éireann came into play there was good confidence that we knew how the storm would play out.

Because the storm was due to hit early on Monday morning for many parts of the country (during the morning rush hour, in fact), warnings were already in place from the Friday before so that schools, businesses and the public would have time to prepare. The heaviest rain associated with any hurricane comes to the north and west of the centre of the eye of the storm, and with Ophelia forecast to track along our west coast, thankfully much of the very heavy rain associated with Ophelia was expected to go out to sea.

The biggest threat to life associated with many of the larger hurricanes comes from the tidal surges created by the strong winds – thankfully these also weren't going to be as big a danger for Ireland because of the orientation of the winds, the length of time the hurricane (or rather ex-hurricane) would be sending those strong winds over the oceans in our direction, and the length of the fetch – the distance travelled by the waves or ocean swell.

The strongest wind ever reported in Ireland happened when Ophelia reached Fastnet Rock off the coast of Cork. Winds gusting up to 191km/h were recorded there early on Monday morning. A buoy off Hook Head, the one that regularly appears on the sea area forecast as the 'Buoy M5', was knocked out of operation during the storm, and the climate station at Sherkin was damaged early in the day.

A strong, dry, persistent southeast wind swept through the whole country first as the winds revolved anticlockwise around the centre of the eye of the storm. Later on, the winds would veer to the southwest as the storm moved up along the coast. People in Kerry reported clear blue skies and dead calm conditions as the eye of the storm passed overhead. Heavy rain pushed over the north and west as the storm moved northwards later in the day, with Donegal reporting areas of spot flooding in the heavy downpours.

The first report of a casualty came early in the day when it was reported a woman in Waterford lost her life when the car in which she was travelling was hit by a falling tree. Later a man died in similar circumstances close to Dundalk in Co. Louth. The third fatality directly attributable to Ophelia happened when a man attempting to clear a tree died in tragic circumstances during the storm.

Trees and power lines were knocked down all over the country, causing more than 360,000 homes to be without power. The roof of a stadium in Cork collapsed, and the roof of a school was recorded flying through the air. Thousands of trees were lost across the country and many buildings were damaged. Although wind speeds recorded at Dublin were not as high as in other storms, as many as 70 mature trees across the city nevertheless came down – one a very large one on Leeson Street. Leeson Street in Dublin is a busy thoroughfare, and on any other Monday would have been bustling with its usual bumper-to-bumper traffic and the constant movement of pedestrians. The red warning in place meant the street was practically deserted. Lives were undoubtedly saved.

In the aftermath of the storm, gardeners across the country noticed that many plants had been destroyed by the strong, dry wind that swept through the country.

The leaves were effectively burned dry. The effect was noticeable, particularly in larger trees, where there was an almost geometric line dividing the burned leaves facing the southeast wind and the leaves on the other side of the trees.

People in the UK and across Europe described an eerie orange sky and reported that the air seemed filled with dust. The winds of Hurricane Ophelia had swept sands from as far away as the Sahara Desert in Africa, all the way over the whole of Europe. The strong, dry wind caused devastation in Portugal as it spread wildfires throughout the country. Throughout the day, Dublin residents reported that the same huge rainbow hung in the sky for almost the entire day.

EPILOGUE

L ike almost all things in life, this book developed
from a vague idea. The sea area forecast is the part
of my job I love the most, because the preparation and
production of it is, essentially, the essence of the job.
There is a lyrical – almost romantic – aspect to the
broadcast of the forecast, especially on national radio
when it either wakes the nation up in the morning or
puts it to bed at night. Its unchanging and predictable
nature assures those that listen to the forecast that they
will always get what they expect. They like and appreciate that pattern, even if they might not particularly *use*
the sea area forecast or even understand what they are
listening to at times. Now that you've read it (assuming
that you haven't just skipped to the end!), I hope I've gone
some way towards bringing you up to speed on what it
all means when we say winds backing north, or pressure
... falling slowly.

But I could have done that with a pamphlet or a
blog post or a link to the Met Éireann website, where
a huge amount of information is available for anyone
to plough through if they like. (Really, it is a fantastic resource, and I wouldn't be doing you a full

service if I didn't advise you to go have a wander –
it's www.met.ie.)

But I hope that this book has given you something
more than that. There is a romance and remoteness to the
headlands and far-flung corners of this beautiful island
that is captured in the sea area forecast and I wanted
to see that for myself and take you, the reader, on that
journey with me. I hope I've managed to combine my
love of science – and, more specifically, my love of
communicating that science – with my love of place and
this island in a way that takes you out of the everyday
and into the magic and mystery of our weather and the
role it plays in our daily lives.

Being married to a Dutchman, I've spent the last
20 years travelling to the Netherlands to visit his family
during any holidays. This was and is the right thing to
do – we gave our children access to their family there.
However, we missed out on Ireland as a consequence.
I realised, suddenly, that my children hadn't ever been
to Galway! We've been to practically every town in the
Netherlands – and that's been great – but I hadn't been
to West Cork since 1980. That's *not* great.

While I was thinking all this through, it was
coming up on the one-year anniversary of the death of
my mother. In 2021, I was devastated when I lost her
suddenly following a routine minor surgery. I had experienced loss before, and I knew well the wisdom we all
share with those who have suffered the pain of losing

a loved one: 'The first year is the hardest.' And so I got through the first year. But as the anniversary approached, I realised that I was almost holding my breath, waiting for the first year to pass so that things would get easier – so that I could move on to the part that was going to be not the hardest.

As the day approached, I had a building tension in my head and body, and I felt as though I needed to run away. I looked to the west and got in my car and drove to Loop Head. The clearing, bright blue skies as I drove out of the rain and into the sun were like a metaphor for the end of the first year of my grief at the loss of my dear mother: moving from darkness to light, from rain to sun.

That journey was an incredible success and I had one of the best days of my life. I cleared my head; I cleared my humour. I came home energised and excited and enthusiastic about writing this book. The next months, as I went from headland to headland, were among the most rewarding of my working life. Travelling mostly alone gave me the opportunity to meet and talk to locals and enjoy that one thing that Irish people do more than anyone else in the world – chat.

I was accosted on Valentia Island by the local school principal and was delighted to visit the schoolchildren there to have a chat about the weather and what we do in the forecasting office. I regularly visit schools around the country (remote visits have been handy since 2020 brought its surprise change to how we live!), and this developing part of my role – engaging with children from an early age to encourage interest in STEM (science, technology, engineering and maths) – has opened up a rewarding and fun avenue in my career. So the opportunity to visit the kids while I was in Valentia was very much welcomed.

My visit to Schull, where I caught up with the O'Driscoll family, opened my eyes to just how exposed the towns and villages along the south and west coasts are – and, more important, how exposed the people living there are – to the gales and storms that crash in off the Atlantic. We tend to visit coastal towns and villages on holiday, and we tend to take holidays more in the summer than in the winter – and even if we do take a winter holiday, we might end up spending more time indoors than out. So my chances to experience onshore gales have been limited. Standing there on the coast and watching the weather roll in, and chatting with these people, whose lives literally depend on the weather forecast, was a great professional bonus to me.

In the north, I talked to locals about the devastating impact the mica issue has had on the community, and I saw the damage to homes dotted along the peninsula of Inishowen. Mica is a material found in concrete, and when too much mica is allowed into the mix it can weaken the blocks. Mica absorbs water, and when the water freezes and thaws in extreme conditions this can significantly exacerbate the weakening process. The abnormally cold winter of 2009/2010 – the coldest in 50 years – followed quickly by another very cold spell in November and December of 2010, is thought to have played a role in the mica problem that first started to emerge in the area during that time.

Each visit was an education and led me to want to see and learn more. But I didn't get everywhere. I'd been to Carnsore Point and Hook Head and had to make a choice of which headland to include along the south coast – much as I do when deciding on the split in the weather in the sea area forecast, although for different reasons. So, I have to go again, and this time visit the ones I missed: Dungarvan and Rossan Point.

Dungarvan is a beautiful, historic harbour town. It dates back to the time of the Vikings, and a castle commissioned by King John stands at the harbour. The Waterford Greenway is a fantastic amenity that I can't wait to check out with my family, so I'm saving my visit to the sunny southeast for a time when I can sit back, relax and enjoy the area fully.

Rossan Point is at the far end of the country from Dungarvan, and – although admittedly I say this a lot! – it genuinely is one of my favourite parts of the country. I spent my childhood summers at the Gaeltacht in Donegal, in a small town called Doochary, and from there, our *múinteoir* Peadar Ó Ceallaigh took us on tours of the nearby county (Donegal is huge, by the way). It was Peadar who first introduced me to the best beach in the world – in my humble opinion – Naran, or Portnoo. With sand like salt and crystal-clear water, it's yet to be matched in my esteem. Rossan Point is south of Naran on the headland that is the furthest west in Donegal. I'm saving that one up for later, too.

I hope that by combining these visits with my love of the weather – and the science of the weather – I have given you an insight into what I live and breathe. I also hope that my amateur hiking skills might give some an incentive to follow in my footsteps.

The evolution of the weather forecast over the past 170 years or so has undoubtedly saved many lives. As technology advances, the accuracy of the short-range forecast has become exceptional. The sea area forecast covers 24 hours with an outlook for a further 24, and with the current high-resolution models available, the service provided to mariners off our shores will continue to save lives in the future. The availability of satellite, radar and direct model output added to the forecast means that seafarers are more informed than ever before.

But while they have a radio, they'll still have our voices, to guide them through our waters – and away from danger.

ACKNOWLEDGEMENTS

This work would not have been possible without the support and input of an abundance of people. First of all, I want to acknowledge the contribution of my husband, Harm Luijkx. Harm is not only my best friend and partner, he is also one of my favourite and most respected colleagues at Met Éireann. He has been invaluable in helping to tease out the science behind the weather events described in the book in the effort to translate complicated atmospheric physics into a language that is accessible to all. As my colleague, our frequent conversations about weather and atmospheric science never cease to educate me. As my husband and partner, his willingness to take on the full burden of the responsibilities of our home and family while I took the time to travel around the country helped make this little book a reality.

I also acknowledge the contributions of those I met along the way who shared their experiences of life at or near the headlands. For me, meeting people is always the high point of any journey. I was incredibly fortunate to happen upon, most often by complete chance, some of the most obliging, friendly and interesting people on our island.

I have a lot of thanks to give, in no particular order of merit. Many thanks to Peter and Teresa Pane, who I met on the very first leg of my journey and who set the tone of the experience for me right away. They were friendly and open, and had that quintessential invaluable trait of the Irish, the willingness to tell stories to a stranger. Thank you to Martin and Simon, the guides at the lighthouse on Loop Head. The great lighthouses of Ireland are an absolute must-see and I was so fortunate to start at one of the most spectacular in the world. Martin and Simon can tell you everything you need to know should you take my advice and go there. When Councillor Martin McDermott heard I was coming to visit Inishowen, he volunteered the hugely charismatic Ali Farren to help show me around. I felt like I'd made a good friend by the end of our afternoon together, we spent so long chatting in the bar, we nearly missed visiting the headland as the short December evening closed in around us. When I visited Hook Head, I chatted with Liam Ryan and he provided me with stories of life as a lighthouse keeper. Jon Pearse is the guide at the lighthouse at Hook Head. There are more guides, but Jon Pearse is THE guide. If you visit, make sure you have time to take full advantage of this man's stories. At Mizen Head, I got to catch up with my friend Councillor Caroline Cronin and the O'Driscoll family – a big thank you for their time and the lovely afternoon of hospitality in gale-force winds and driving freezing rain. One of the last visits I made was to Valentia Island and I was most fortunate to meet with

Father Patsy Lynch, who gave me enough information to fill a whole separate book on Valentia Island alone.

Information in relation to the synoptic situation of the storms mentioned, data related to weather events on the island of Ireland and related climate information is available at met.ie. The Met Éireann website is a treasure trove of information that is available to all. I strongly encourage readers to take advantage of the huge amount of work done by my colleagues at Met Éireann in collating the information and making it publicly available. Take some time to explore and enjoy!

I would like to thank the group at Gill Books for all their support along the way. Particularly to Teresa Daly, who nudged me into the start of this project that ended up being so much more meaningful to me than she could ever understand. The helpfulness of the editorial work of Rachel Thompson and Margaret Farrelly cannot be understated.

INDEX